Raspberry Pi® Projects for the Evil Genius™

Evil Genius™ Series

Arduino + Android Projects for the Evil Genius
Bike, Scooter, and Chopper Projects for the Evil Genius
Bionics for the Evil Genius: 25 Build-It-Yourself Projects
Electronic Circuits for the Evil Genius, Second Edition: 64 Lessons with Projects
Electronic Gadgets for the Evil Genius, Second Edition
Electronic Gadgets for the Evil Genius: 28 Build-It-Yourself Projects
Electronic Sensors for the Evil Genius: 54 Electrifying Projects
15 Dangerously Mad Projects for the Evil Genius
50 Awesome Auto Projects for the Evil Genius
50 Green Projects for the Evil Genius
50 Model Rocket Projects for the Evil Genius
51 High-Tech Practical Jokes for the Evil Genius
46 Science Fair Projects for the Evil Genius
Fuel Cell Projects for the Evil Genius
Holography Projects for the Evil Genius
Mechatronics for the Evil Genius: 25 Build-It-Yourself Projects
Mind Performance Projects for the Evil Genius: 19 Brain-Bending Bio Hacks
MORE Electronic Gadgets for the Evil Genius: 40 NEW Build-It-Yourself Projects
101 Outer Space Projects for the Evil Genius
101 Spy Gadgets for the Evil Genius, Second Edition
123 PIC® Microcontroller Experiments for the Evil Genius
123 Robotics Experiments for the Evil Genius
125 Physics Projects for the Evil Genius
PC Mods for the Evil Genius: 25 Custom Builds to Turbocharge Your Computer
PICAXE Microcontroller Projects for the Evil Genius
Programming Video Games for the Evil Genius
Raspberry Pi Projects for the Evil Genius
Recycling Projects for the Evil Genius
Solar Energy Projects for the Evil Genius
Telephone Projects for the Evil Genius
30 Arduino Projects for the Evil Genius, Second Edition
tinyAVR Microcontroller Projects for the Evil Genius
22 Radio and Receiver Projects for the Evil Genius
25 Home Automation Projects for the Evil Genius

Raspberry Pi® Projects for the Evil Genius™

Donald Norris

New York Chicago San Francisco Lisbon London Madrid
Mexico City Milan New Delhi San Juan Seoul
Singapore Sydney Toronto

McGraw-Hill Education books are available at special quantity discounts to use as premiums and sales promotions or for use in corporate training programs. To contact a representative, please visit the Contact Us page at www.mhprofessional.com.

Raspberry Pi® Projects for the Evil Genius™

2 3 4 5 6 7 8 9 0 QVS/QVS 1 9 8 7 6 5 4 3

ISBN 978-0-07-182158-2
MHID 0-07-182158-9

This book is printed on acid-free paper.

Sponsoring Editor
Roger Stewart

Editing Supervisor
Stephen M. Smith

Production Supervisor
Pamela A. Pelton

Acquisitions Coordinator
Amy Stonebraker

Project Manager
Nancy Dimitry,
D&P Editorial Services

Copy Editors
Joe Cavanagh,
Nancy Dimitry,
D&P Editorial Services

Proofreader
Don Dimitry,
D&P Editorial Services

Indexer
WordCo Indexing Services

Art Director, Cover
Jeff Weeks

Composition
D&P Editorial Services

To Karen, my lovely soul mate and most ardent supporter.
Her faith in me never wavers and is my core strength.

About the Author

Donald Norris has a degree in electrical engineering and an MBA specializing in production management. He is currently teaching undergrad and grad courses in the IT subject area at Southern New Hampshire University. He has also created and taught several robotics courses there. He has over 30 years of teaching experience as an adjunct professor at a variety of colleges and universities.

Mr. Norris retired from civilian government service with the U.S. Navy, where he specialized in acoustics related to nuclear submarines and associated advanced digital signal processing. Since then, he has spent more than 17 years as a professional software developer using C, C#, C++, Python, and Java, as well as 5 years as a certified IT security consultant.

Mr. Norris started a consultancy, Norris Embedded Software Solutions (dba NESS LLC), that specializes in developing application solutions using microprocessors and microcontrollers. He likes to think of himself as a perpetual hobbyist and geek and is always trying out new approaches and out-of-the-box experiments. He is a licensed private pilot, photography buff, amateur radio operator, avid runner, and, last but very important, new grandfather—here's to you, Hudson.

Contents

Preface

This Raspberry Pi project book is not only about building a series of interesting projects but also about providing an education regarding the underlying project technologies. I am positive that my over-30-years' experience as a college educator forced me to ensure that readers could not only build the projects but also understand why they function as designed.

Building a successful project is rewarding unto itself, but understanding why and how it functions is far more important. The reader should expect a manifold increase in experience with the Raspberry Pi if a commitment is made to expend the time and energy to complete most, if not all, of the projects. I learned a lot while completing them; sometimes things worked out fine, while other times they were not successful. But that's the joy of experimenting. As Professor Einstein once stated, "Anyone who has never made a mistake has never tried anything new."

The joy of learning about and building projects is the core concept within this book. I designed and built all of the projects, and along the way gained a lot of knowledge about the Linux OS and how it really shines as an embedded development platform.

I will not recap the projects here other than to state that the complexity increases from the book's beginning to its end. And this is how it should be, as experience and confidence in dealing with Linux and the Python language are progressively gained by proceeding through the projects.

Experienced Linux developers should feel free to jump into any of the projects; however, there are useful hints and techniques sprinkled throughout the book, which might be missed by taking a selective approach to reading it. I have also tried to point out the constraints and limitations of the Raspberry Pi as I encountered them when designing and building the projects. Just keep in mind, a $35 computer simply cannot meet all expectations.

One disclaimer that I feel is warranted relates to the Python programs. These programs, while fully functional for the respective projects, are probably not in their best form. As I tell my beginning programming students, there are many ways to develop functional programs. Some are better than others—not necessarily right or wrong. With this perspective, I tried to keep the programs simple and to the point, and to avoid any unneeded complexity.

All of the book's projects can be expanded and modified as desired. I strongly recommend that readers do so, as that is one sure way to understand the concepts and bolster skills. The ability to experiment has been described as one of the key attributes that modern employers are looking for in twenty-first century employees.

Donald Norris

Acknowledgments

I thank Karen for putting up with all my experiments and enduring all the "discussions" about the book's projects.

I also thank Roger Stewart for his continued support and guidance as the sponsoring editor. He is the true champion for this book.

Thanks to Amy Stonebraker for her support as editorial assistant.

Thanks also goes out to Nancy Dimitry for her fine work as the project manager.

Finally, I would like to thank all the folks at the Raspberry Pi Foundation for creating the board and getting it to the marketplace.

Raspberry Pi®
Projects
for the
Evil Genius™

CHAPTER 1

Introduction to the Raspberry Pi

THIS BOOK WILL INTRODUCE you to the Raspberry Pi and provide 12 projects that will lead you through some simple, fundamental operations up to some fairly complex ones. The Raspberry Pi, despite its small size, is a fully functional computer capable of running a full-fledged Linux operating system. It is also the most inexpensive computer with this level of functionality that is presently available to the public.

The Raspberry Pi is a small board measuring 56 by 85 mm, about the size of a standard credit card. Nonetheless, it contains some very impressive functionality, as you will discover later in this chapter. This new board is shown in Fig. 1-1.

A look at a bit of history regarding the Raspberry Pi (or RasPi as I will now refer to it) and its originally intended market may help you understand the constraints and limitations that subsequently ensued. The RasPi concept began around 2006 with Dr. Eben Upton and his colleagues at the University of Cambridge's Computer Laboratory in Cambridge, England. They were concerned about the decline in knowledge and skill levels of incoming computer science students as compared with those of earlier students. Dr. Upton decided to create an inexpensive computer, reasoning that it was likely that parents were not allowing their children to experiment with modern and relatively expensive PCs. This idea ultimately led to the development of the very inexpensive RasPi. This computer would provide an excellent opportunity for children to learn and experiment with programming, while not being a concern to parents if something should go horribly wrong and the board be destroyed.

Dr. Upton teamed with several other individuals to form the Raspberry Pi Foundation, a registered United Kingdom charity that promotes computer literacy and enthusiasm, especially among young children using the RasPi as their initial platform. They seem to be achieving these highly laudable goals, since they have greatly exceeded the initial estimate of selling 10,000 RasPi's, and at the time of this writing, the total sales are approaching one million units. The foundation's website is www.raspberrypi.org, where you will find all sorts of information about the board, current news, forums, FAQs, and so on.

A key design decision that kept costs low was to incorporate a SoC type chip on the board. SoC is short for *System on a Chip*—a technology that physically places the memory, microprocessor, and graphics processer in a type of silicon "sandwich" that in turn minimizes the *printed circuit board* (PCB) space and the accompanying PCB interconnecting board traces. The foundation eventually partnered with Broadcom to use its designs for both the microprocessor and graphics processors in the SoC. The SoC and some other key components and connections that you should know about are identified in Fig. 1-2.

Figure 1-1 The Raspberry Pi, a small Linux computer.

Although it is not critical to understand the Broadcom microprocessor in order to use the RasPi, it is still helpful to discuss it for a bit so that you will know why the RasPi is slower than your PC and why the low voltage of 3.3 V is used for interfacing the RasPi to the outside world. I will first cover the hardware aspects of the RasPi, followed by the software aspect.

Hardware

Broadcom 2835 Microprocessor/ Graphics Processing Unit

The SoC uses the Broadcom BCM2835 as its microprocessor and *graphics processing unit* or GPU. The Broadcom company is what is known as a *fabless* supplier in that they provide the designs for their product in the form of *Intellectual Property* (IP) and other companies actually create the real silicon chips. Broadcom specializes in mobile-application-type processors including the type used in smartphones. The BCM2835 portion of the SoC itself is made up of an ARM1176JZF-S microprocessor running at 700 MHz and a Broadcom VideoCore® IV GPU.

The BCM2835 is designed for mobile applications, and hence, it needs to operate with minimal power so as to extend battery life. A fairly low microprocessor clock speed helps lower power consumption, and this is the reason the

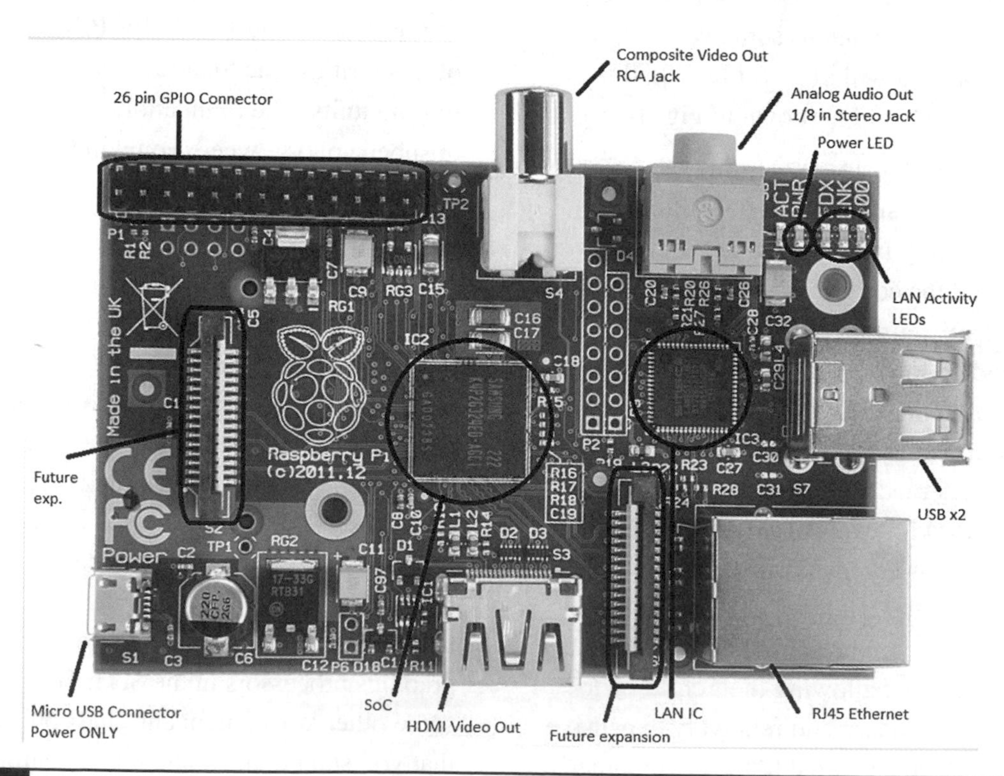

Figure 1-2 The SoC and other key components.

BCM2835 operates at 700 MHz, which is typically a quarter of the speed of a modern PC. Lower clock speed also means the processor can operate at a low voltage, thus decreasing the overall heat generated and extending chip life. The BCM2835 can be speeded up—also known as overclocking—to improve performance, but this is generally not recommended because the microprocessor can become operationally unstable and its life shortened. Be assured that the RasPi is sufficiently fast for all the projects in this book.

Broadcom has also graciously provided software drivers to allow the BCM2835 input and output pins to be connected to external peripherals. This software is in the form of a Python library that I will discuss later.

The Broadcom VideoCore IV GPU handles all the video and audio processing for the SoC. This GPU directly supports the OpenGL ES 2.0 standard that is essentially an *Application Program Interface* (API) capable of running on embedded hardware, which, in this case, is the Broadcom 2835. Loosely translated, this means the 2835 can easily display three-dimensional 3D graphics using all the requisite shaders and texture filters normally required for modern games and *high-definition* (HD) video. This chip implements in hardware a 1080p, 30 frames/sec, H.264 codec required for HD. That is an impressive performance.

I will not pursue this discussion any further other than to state that the BCM2835 is more than adequate to display all the graphics and output all the audio streams required for all the projects in this book.

For readers fascinated with performance statistics, it is interesting to note that the Broadcom Videocore IV GPU has the following processing capabilities:

- 1 gigapixel/sec (that's one billion pixels processed per second)
- 1.5 gigatexels/sec (that's one and a half billion texture elements per second)
- 24 gigaflops (that's 24 billion floating point operations per second)

All of this capability translates to the equivalent performance of a first generation Xbox®, not bad for a small embedded chip in a SoC sandwich!

Memory

There are two memory types used in the RasPi: *dynamic random access memory* (DRAM) and *Secure Digital* (SD) flash. The original version, model A, of the RasPi had 256 MB of RAM installed, while the latest, Model B, has 512 MB. The 512-MB chip is easily seen on the board as the top layer of the SoC sandwich. The SoC chip shown in Fig. 1-2 has top DRAM marked as supplied by Samsung with a part number K4P4G324EB ACC1, which translates to a low-power 4-Gbit (512-MB) DRAM designed for mobile applications. This means that it too uses low voltage while maintaining reasonable clock speed. Having 512 MB of DRAM means the operating system will function very efficiently and programs should also run smoothly provided they are properly created.

The SD flash memory is used to store the operating system, all programs, and all other data that need persistence. In other words, nothing will be destroyed when the power is shut off. The RasPi uses SD flash memory in the same manner that a PC uses a hard drive to permanently store data and programs. You have a choice in selecting the memory size of a SD memory card that simply slides into a holder that is located on the underside of the RasPi, as shown in Fig. 1-3.

If you purchased your RasPi as part of a starter kit, you will have received a 4-GB SD card with a Linux OS distribution already installed on the card. A picture of this preprogrammed SD card is shown in Fig. 1-4.

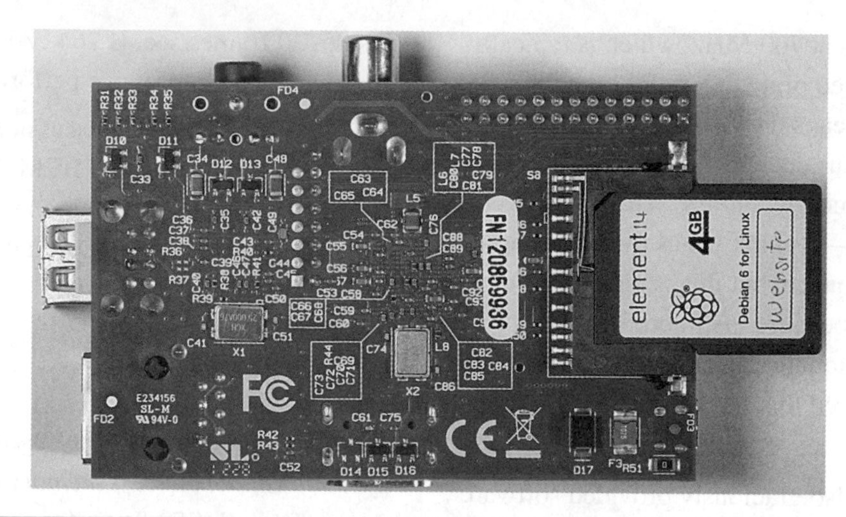

Figure 1-3 Back side of the Raspberry Pi.

I will explain later on in this chapter how to create your own memory card so you do not have to purchase a preprogrammed SD card. Note that most SD cards also have a class designation on the label. The one shown in Fig. 1-3 does not, as it has had a customized label attached. The class designation relates to the minimum data-transfer speed the SD card can handle. Remember, the SD card is taking the place of a hard drive, so the faster, the better. Class 4 is very typical of the consumer grade SD cards that are available in most office supply stores. Class designations and their associated minimum transfer speeds are shown in Table 1-1.

Figure 1-4 Preprogrammed SD card.

What you should take away from this SD class discussion is that the higher the class number of the SD card used in the RasPi, the better it will perform. The only downside is that SD cards with high class numbers are more expensive than ones with lower numbers, sometimes more than twice the cost for the same storage capacity. My only suggestion is to purchase a class 4 or higher; anything less and you will be disappointed in your RasPi's slow response.

RasPi Connectors

The RasPi has nine connectors: power, *High-Definition Multimedia Interface* (HDMI), analog composite video, audio, ethernet, *Universal Serial Bus* (USB), two future expansion connectors,

Table 1-1 SD Card Class Designations

Class	Minimum Performance
Class 2	2 MB/sec
Class 4	4 MB/sec
Class 6	6 MB/sec
Class 10	10 MB/sec

and the *General Purpose Input/Output* (GPIO) interface. Each connector has specific functions that I will discuss in the following sections, except for the expansion connectors, which are not yet used, although I will tell you what I know about them as of this writing. There is no particular order to this discussion although I have left the GPIO connector for last because it is by far the most complex and, hence, requires the most explanation.

Power Connector

The power connector shown in Fig. 1-5 is a micro USB socket that is wired to pass the 5-volt (V) *direct current* (DC) lines from a micro USB plug, also shown in the figure. (Since all voltage in this project is DC, I will use just the notation V for V DC.) No data connections are wired to this socket. You can use almost any available smartphone charger that has a micro USB connector or use the power supply that came with the RasPi kit, if that's what you purchased.

Figure 1-6 shows a RasPi kit power supply that is rated to supply 5 V at 1000 *milliamperes* (mA) or 1 *ampere* (A). The regulatory compliance

Figure 1-5 Micro USB power connector.

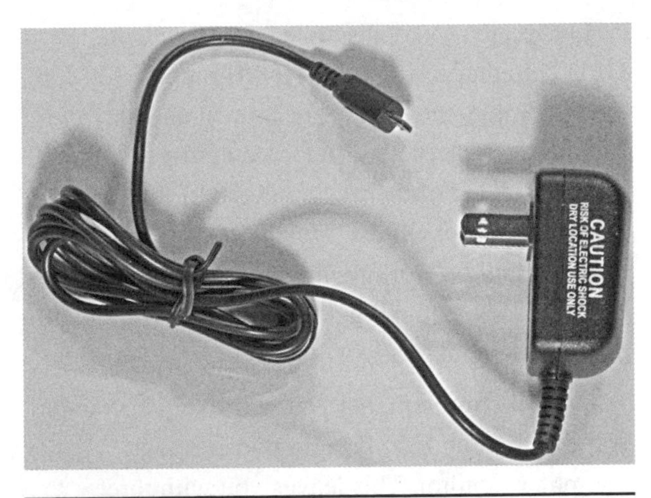

Figure 1-6 External power supply.

document supplied with this RasPi states the following:

> This product shall only be connected to an external power supply rated at 5 V, and a minimum current of 500–700 mA for model A and 700–1200 mA for model B.

I will have a bit more to say regarding current consumption when I discuss the USB connector.

HDMI Connector

The RasPi provides video and audio using a fully compliant HDMI, which is modern by most standards The board socket and sample cable plug are shown in Fig. 1-7.

Figure 1-7 HDMI connector and cable.

I have previously discussed the Broadcom GPU chip that controls the HDMI output. To keep things simple, the book projects will use only the "standard" type of audio/video output and will not take advantage of the true potential of the RasPi's multimedia capabilities. Trust a fellow Evil Genius that you will be working hard to complete the book projects without getting involved with HDMI development tasks.

One real problem that you will likely encounter is the lack of an HDMI input port for your computer monitor. This leaves you with three choices for observing the RasPi video:

1. Use the composite video output with a compatible analog monitor
2. Use an adapter to convert from HDMI to *Video Graphic Array* (VGA) or *Digital Video Interface* (DVI)
3. Take over the family's flat panel digital TV

The first option is really not a very good choice, since the quality is diminished as compared to what is displayed by a high-quality computer monitor. The second option is the preferred method, as it yields the best results using your existing computer resources. Choosing the third and final option will likely result in family discord and upheaval for which I will take no responsibility!

The choice of an HDMI to VGA or HDMI to DVI adapter will, of course, depend upon what type of monitor input you have. Most monitors have a VGA input, and an adapter for that type of input is shown in Fig. 1-8. The HDMI to DVI adapter is similar, and the cost for each is also similar.

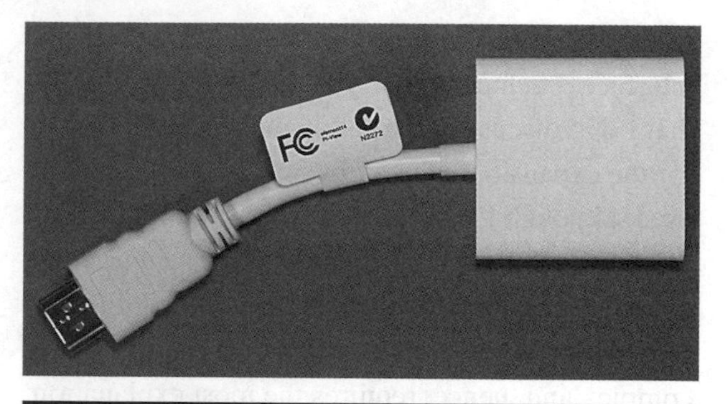

Figure 1-8 HDMI to VGA adapter.

The HDMI connection also contains a very interesting surprise. The RasPi can act as a very sophisticated remote control for HDMI-CEC compliant devices. The CEC suffix is short for *Consumer Electronics Control*, which is a one-wire, bidirectional serial bus link protocol used for the control of audio and video devices. HDMI-CEC has been implemented by many A/V manufacturers including Sony with its Bravialink, LG with its Simplink, Sharp with its Aquos Link, Samsung with its Anynet+, and so on. The bad news is that there is currently no RasPi software support available for HDMI-CEC remote control functions. The good news is to simply wait for a short time because new software apps are constantly being created, free of charge. By the time you are reading this, the RasPi should be able to turn your digital flat-panel TV on and off as well as control your A/V receiver, DVD player, Blu-Ray player, etc. The RasPi will become the ultimate remote control. For more information, go to http://elinux.org/CEC_(Consumer _Electronics_Control)_over_HDMI.

Analog Composite Video Connector

The RasPi also produces an analog video output from the RCA socket, as shown in Fig. 1-9.

Figure 1-9 Analog video connector and cable.

This analog video functionality was deliberately included in the RasPi design to accommodate all those situations where only analog monitors or analog TVs are available, especially in developing countries. There is, however, an upside to the composite output. To monitor project parameters in real time, you can use small analog monitors. These monitors are fairly inexpensive and can often be battery powered, which is not a realistic option with larger computer monitors. I have included the use of a small, battery-powered analog monitor in one of the book projects. This monitor is shown in Fig. 1-10.

Audio Connector

The RasPi is also capable of creating an analog audio output in full stereo. The output is from a standard 3.5-mm stereo jack as shown in Fig. 1-11.

This audio would normally be the analog equivalent of the digital audio outputted from the HDMI connector. There is a book project that uses this analog output to play MP3 songs. You will need an audio amplifier to hear the music, as the RasPi does not generate a powerful enough signal to drive an unamplified speaker. However, a good quality set of headphones will work.

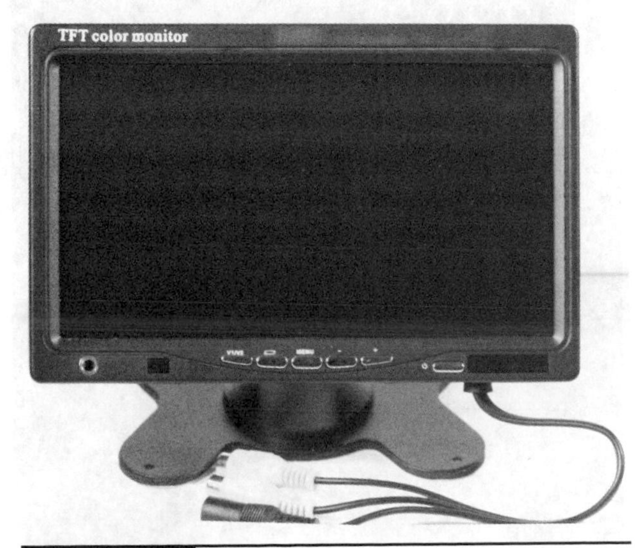

Figure 1-10 Small analog video monitor.

Figure 1-11 Analog audio connector and cable.

Ethernet and USB Connectors

Both the Ethernet and USB connectors are shown in Fig. 1-12. I will discuss the Ethernet connector first, followed by the USB connectors.

The Ethernet connector shown on the left in the figure is a standard RJ45 connector. You would

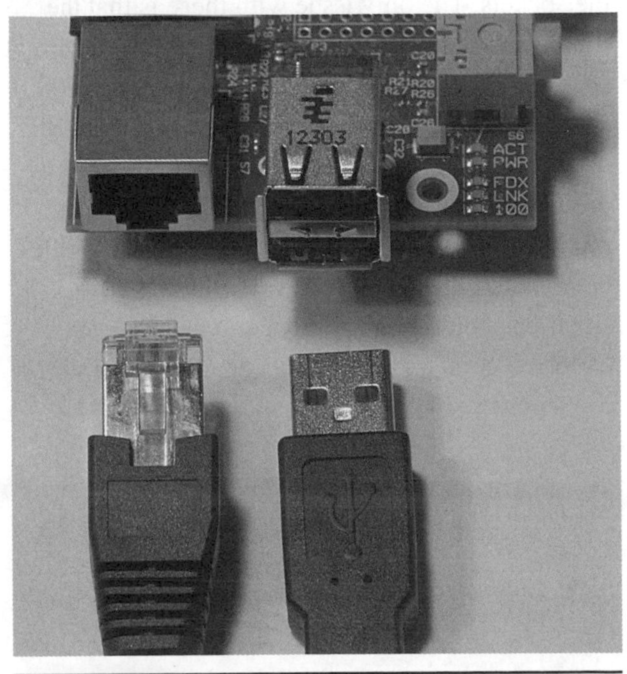

Figure 1-12 Ethernet and USB connectors with cables.

simply plug your standard Ethernet patch cable into the socket and connect the other end to either your router or switch, if that is the way you have setup your home network. The RasPi will then automatically "negotiate" with your router to gain an *Internet Protocol* (IP) address in a process known as *Dynamic Host Configuration Protocol* (DHCP). There are five *light-emitting diodes* (LEDs) to the left side of this socket as you look at it head on. The LED furthest away from the socket is labeled "100." If it is shining with a yellow light, this means that a 100-*megabits-per-second* (Mb/s) connection was made. The next two LEDs, just to the right of the 100 LED, are labeled "LNK" and "FDX". These LEDs shine with green lights to indicate that the Ethernet is alive and operating. Checking these LEDs is a quick way to determine if your Ethernet connection is working or if something, somewhere, has gone down.

There is a stack of two USB connectors shown on the right-hand side of the figure. These are normal USB connectors in the sense that USB peripherals will be recognized when plugged into the sockets. The only issue with these is that the RasPi cannot supply the standard amount of current according to the USB specification, which is 500 mA per socket. Remember that I mentioned earlier in Fig. 1-6 that the power supply in the RasPi kit provides up to 1000 mA. If peripherals plugged into these sockets took 500 mA each, there would be none left for the poor RasPi! Obviously, this situation should not be allowed to happen, and there is a good and relatively cheap solution. I use a powered USB hub, as shown in Fig. 1-13,

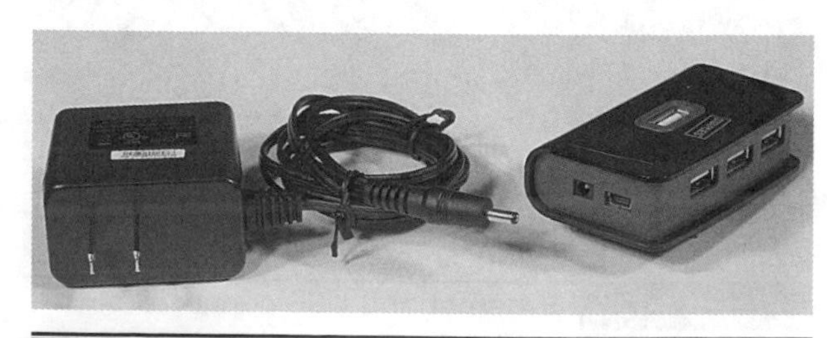

Figure 1-13 Self-powered USB hub.

that can easily provide all the current that typical unpowered USB peripherals require.

There is one USB cable that connects between the hub and the RasPi. That leaves one available USB socket on the RasPi for a low-power peripheral, such as a thumb drive. The number of USB ports provided by the hub varies with the manufacturer; however, four or five ports are fairly common. The power supply shown in the figure is rated for 2100 mA, which precisely matches the USB specification for four ports and a little left over for the hub internal electronics.

Future Expansion Connectors

Two connectors prominently populated on the RasPi are not currently used. Referred to as "future expansion" connectors, they reflect the dynamic nature of the RasPi project. Fig. 1-14 is a close photo of one of the connectors, labeled "S2." This connector is a 15-way, flat-flex connector designated for use with the *Camera Serial Interface* (CSI-2). A prototype digital serial camera was just introduced at an international electronics show at the time of

Figure 1-14 Expansion connectors.

this writing. The other flat-flex connector labeled "S5" and located just behind the Ethernet RJ45 connector is designated as a *Display Serial Interface* (DSI) that will eventually drive a Sony *Low Voltage Differential Signaling* (LVDS) serial display. You should check the RasPi website for the latest news regarding the RasPi.

GPIO Pin Interface Connector

The General Purpose Input Output (GPIO) connector has 26 pins positioned in two rows of 13 pins each. Fig. 1-15 shows this connector with pins 0, 1, 25, and 26 pointed out.

Table 1-2 details pin assignments with both the RasPi pin designations and the BMC2835 pin designations. Using two different sets of pin designations is confusing, but unfortunately, that is the situation with this board. I will try to use the RasPi pin designations whenever possible; however, there will be situations where the software will require the use of the BMC2835 pin designations. I

Figure 1-15 GPIO connector.

will try to be as clear as possible regarding the exact pin that is being used and for what purpose.

The *Universal Asynchronous Receiver/Transmitter* (UART), *Serial Peripheral Interface* (SPI), and *Inter-Integrated Circuit* (I^2C) functional pins listed in the table may all be reconfigured as GPIO pins. These are shown with an asterisk in Table 1-2. This means that up to 17 GPIO pins (8 existing GPIO + 9 reconfigurable) are available for

Table 1-2 GPIO Pin Descriptions

RasPi pin #	RasPi Description	BMC Description	RasPi pin #	RasPi Description	BMC Description
1	3V3	3V3	2	5V0	5V0
3	SDA0*	GPIO0	4	5V0	5V0
5	SCL0*	GPIO1	6	GND	GND
7	GPIO_GCLK	GPIO4	8	TXD0*	GPIO14
9	GND	GND	10	RXD0*	GPIO15
11	GPIO_GEN0	GPIO17	12	GPIO_GEN1	GPIO18
13	GPIO_GEN2	GPIO21	14	GND	GND
15	GPIO_GEN3	GPIO22	16	GPIO_GEN4	GPIO23
17	3V3	3V3	18	GPIO_GEN5	GPIO24
19	SPI_MOSI*	GPIO10	20	GND	GND
21	SPI_MISO*	GPIO9	22	GPIO_GEN6	GPIO25
23	SPI_SCLK*	GPIO11	24	SPI_CE0_N*	GPIO8
25	GND	GND	26	SPI_CE1_N*	GPIO7

Legend:

3V3	3.3 V	SPI_...	Serial Peripheral Interface
5V0	5.0 V	TXD0	Serial Transmit
GND	Ground	RXD0	Serial Receive
GPIO_GENn	General Purpose Input Output pin n	*	Can be reconfigured as a GPIO pin

hardware interfacing, provided that the functions mentioned before are not needed.

Figure 1-16 shows all the GPIO connector pins with the BCM2835 pin designations. You should always crosscheck your connections with this figure anytime that you are directly wiring to this connector.

CAUTION **All GPIO voltage levels are 3.3 V, not 5 V tolerant. There is no overvoltage protection incorporated on the RasPi; and if you inadvertently subject a GPIO pin to 5 V, you will wreck your RasPi. I will not take any responsibility for such foolish actions, as you have been warned!**

I have incorporated hardware buffers into projects where 5-V sensors interface to the RasPi, thus avoiding any chance of damaging the RasPi due to an input voltage overload. You must always pay careful attention to how the projects are wired because it is easy to damage the RasPi through an inadvertent construction mistake.

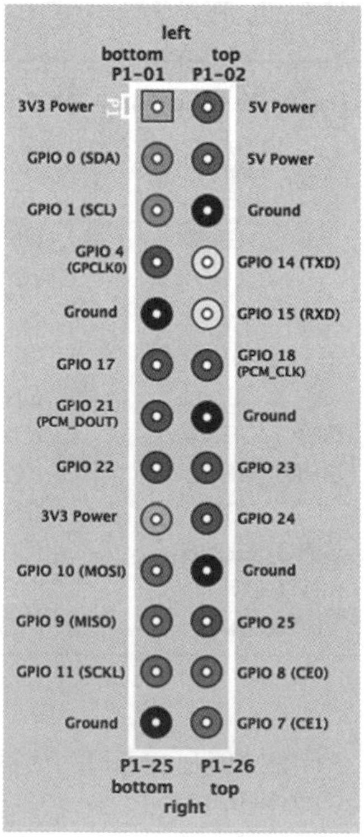

Figure 1-16 GPIO connector pins with the BCM2835 pin designations.

Table 1-3 Digital Signal Voltage Levels

Logic	Voltage Range
1 or HIGH	2.7 to 3.3
0 or LOW	0.0 to 0.7

Digital Signal Voltage Level

The RasPi operates at a *positive power-supply voltage* (V_{dd}) of 3.3 V with the digital logic levels, shown in Table 1-3. This means that any GPIO input voltage greater than 2.7 V will be detected as a logical one or HIGH. Any voltage input that is less than 0.7 V will be detected as a logical zero or LOW. However, the input voltage can never exceed 3.3 V, or it will destroy the GPIO pin circuit.

It turns out that standard 5-V logic accepts 3.3 V as a logical one or HIGH and anything less than 0.7 V as a logical zero or LOW. This is exactly the reason that a RasPi can *output* to a 5-V logical device. The difficulty happens if a 5-V device *inputs* into a GPIO pin. The 5-V logical device has the logic HIGH voltage range of approximately 4.4- to 5-V that will immediately burn out the GPIO pin input circuitry.

Current Limits

There are also some current draw limitations for both the 3.3-V and 5-V power pins. The limitations are dependent upon the RasPi model, as shown in Table 1-4.

Every GPIO pin can sink or source a limited amount of current ranging from 2 mA up to 16 mA. This means that you must be very careful about

Table 1-4 Raspberry Pi Current Consumption vs Model

Raspberry Pi Model	3.3 V	5 V
A	50 ma	1000 – 500 = 500 ma
B	50 ma	1000 – 700 = 300 ma

the current demands put on the RasPi as well as how much current it will accept without causing problems.

GPIO Pin Expansion

Recently, the Raspberry Pi Foundation made a revision to the Model B that added access to some additional GPIO pins that were not available in the earlier production run. This latest board is designated rev 2, while the earlier version is designated rev 1. The additional pins are plated PCB holes, as shown in Fig. 1-17, and are located next to the GPIO connector.

Table 1-5 shows all the additional pins with their RasPi and BMC designations. Pin 1 is the square plated hole located in the upper left corner of P5. You will need to install a 12 pin connector to access the pins. The connector is supposed to be installed on the board's underside per Note 3 on the rev 2.0 board's schematic, which is available at http://www.raspberrypi.org/wp-content/uploads/2012/10/Raspberry-Pi-R2.0-Schematics-Issue2.2_027.pdf. A suggested connector is shown in Fig. 1-18. You will not need any of these additional pins from P5 to build any of the projects in this book.

Figure 1-17 Additional GPIO pins available for expansion.

Figure 1-18 GPIO pin expansion connector.

Interrupts

Each GPIO pin can also accommodate what are known as interrupts. An interrupt is an event that stops or "interrupts" the normal programming flow and directs the microprocessor to execute some special handler program, or code, for the interrupt source. Interrupts may be triggered in several ways:

- HIGH level detected
- LOW level detected
- HIGH to LOW transition detected
- LOW to HIGH transition detected

Table 1-5 Additional GPIO Expansion Pins

RasPi pin #	RasPi Description	BMC Description	RasPi pin #	RasPi Description	BMC Description
1	5V0	5V0	2	3V3	3V3
3	GPIO_GEN7	GPIO28	4	GPIO_GEN8	GPIO29
5	GPIO_GEN9	GPIO30	6	GPIO_GEN10	GPIO31
7	GND	GND	8	GND	GND

Legend:

3V3	3.3 V	GND	Ground
5V0	5.0 V	GPIO_GENn	General Purpose Input Output pin n

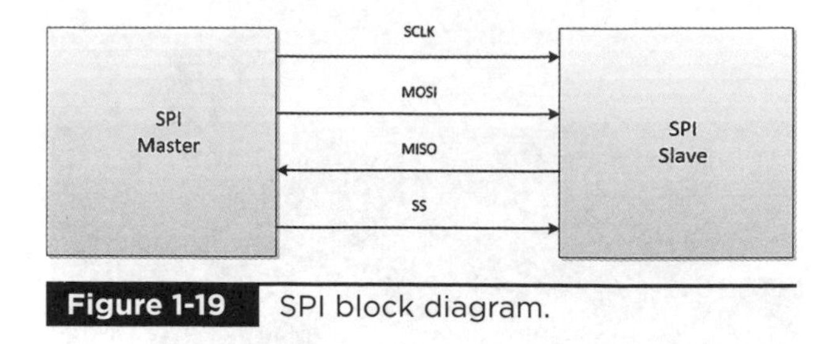

Figure 1-19 SPI block diagram.

Using interrupts will certainly improve performance, but at the expense of adding a certain level of complexity to the software.

Serial Protocols

There are several serial protocols shown in the pin descriptions that I wish to discuss briefly.

SPI Serial Protocol

The first is the *Serial Peripheral Interface* (SPI), which is shown in the Fig. 1-19 block diagram.

The SPI interface (pronounced "spy" or "ess-pee-eye") is a synchronous serial data link. A clock signal is needed because it is synchronous. It is also a full-duplex protocol, which means that data can be simultaneously sent and received from the host and slave. SPI is also referred to as a *Synchronous Serial Interface* (SSI) or a 4-wire serial bus.

The four interconnecting signal lines between the SPI host and SPI slave shown in Fig. 1-19 are explained in Table 1-6.

Table 1-6 SPI Signal Lines

Signal Name	Description	RasPi Name
SCLK	Clock	SPI_SCLK
MOSI	Master Out Slave In	SPI_MOSI
MISO	Master In Slave Out	SPI_MISO
SS	Slave Select	SPI_CEn**

**There are two pin connections on the RasPi named SPI_CE0 and SPI_CE1. These connections allow two separate SPI slave devices to be individually selected or "enabled." In fact, the CE in the RasPi name is short for *Chip Enable*.

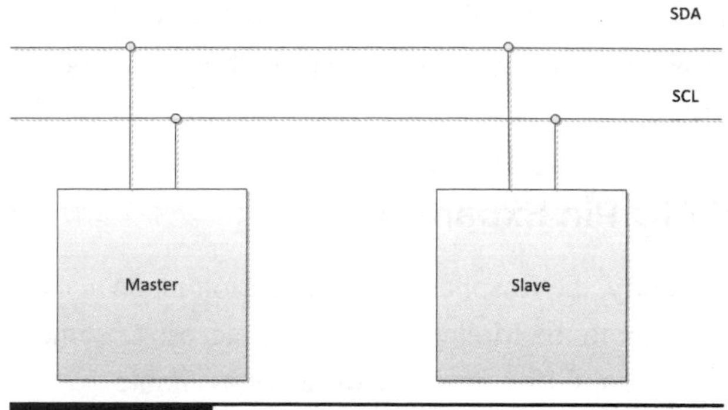

Figure 1-20 I^2C block diagram.

I^2C Serial Protocol

The next serial protocol that I will discuss is the Inter-Integrated Circuit interface or I^2C (pronounced "eye-two-cee" or "eye-squared-cee"), which is also a synchronous serial data link. Fig. 1-20 is a block diagram of the I^2C interface, showing one master and one slave. This configuration is also known as a multidrop or bus network.

I^2C supports more than one master as well as multiple slaves. This protocol was created by the Philips Company in 1982 and is a very mature technology, meaning it is extremely reliable. Only two lines are used: SCLK for serial clock and SDA for serial data. Table 1-7 shows the RasPi names for both the clock and data lines.

UART Serial Protocol

The last serial protocol shown in Table 1-2 is the standard Universal Asynchronous Receiver Transmitter (UART) that uses two pins in the RasPi and is shown in the block diagram in Fig. 1-21.

The UART protocol needs no clock signal, just as it is described by the asynchronous adjective

Table 1-7 I^2C Signal Lines

Signal Name	Description	RasPi Name
SCL	Clock	SCL0
SDA	Data	SDA0

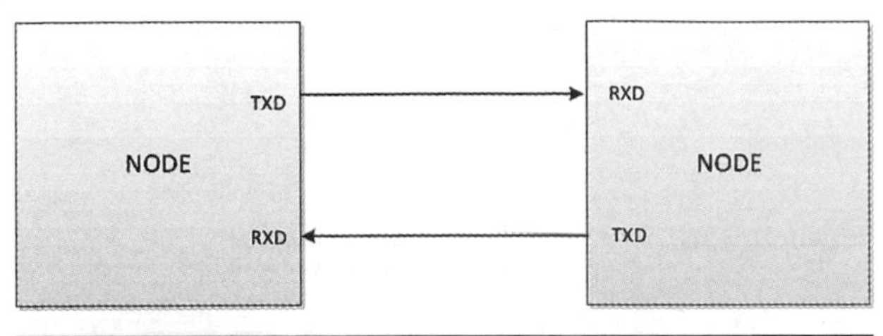

Figure 1-21 UART block diagram.

in its name. The RasPi transmits data on the pin named TXD0 and receives on the pin named RXD0. There is also no concept of a master or slave in this protocol, since it is used primarily for data communications instead of for control, which is the focus of both the SPI and I^2C interfaces.

All three serial protocols described in this section are implemented in hardware, meaning that there is actual silicon dedicated to performing the protocol functions. This is the most efficient and fastest way to provide these serial interfaces, but it is not the only way. You can implement these same serial interfaces by using uncommitted GPIO pins and software. Although this would provide nearly the same functionality as the hardware implementation, it would not be as fast. The term "bit-banging" is often used to describe this approach. Sometimes you must use bit-banging when the hardware is not available.

I will now jump into the heady software arena; I'm sure you already sense that hardware and software are closely linked in the RasPi.

Software

The RasPi was designed to be run with a Linux operating system. This design decision stands in stark contrast to many other similar microprocessor boards, including the popular Arduino series, that do not operate in this fashion. This is not to imply that an Arduino board is inferior to the RasPi but simply to show that using the RasPi brings additional flexibility and capability to projects because of the Linux environment.

I will not start this discussion with a Linux tutorial, since there are many good Linux resources available on the web. Google (or your favorite search engine) will turn up many hits regarding Linux tutorials. I will instead use Linux commands and procedures, and explain them as I go along in a mentor capacity, as if I were standing beside you as you execute the commands. Additionally, I will be using the Python language to program, or code, the RasPi and will provide more guidance regarding how Python works and how it is applied with the RasPi, as it is the key to success in using the board and in understanding its operation with the underlying project code.

Initial Startup

A suggested RasPi setup that uses the connections discussed in the hardware section is shown in Fig. 1-22. This setup will be the basis for your projects once you add some prototyping hardware to the RasPi. Right now, I will be using the setup to get the RasPi configured in a proper manner to enable project development. You should connect all the components as shown in the figure, leaving the USB power connection for last. The RasPi will attempt to start up when the USB power cord is plugged in; and if you have not finished connecting all the other components, it is entirely possible that they will not be recognized in the startup sequence and your system will either not start or not operate correctly. I will also assume that you are using a "prebuilt" Linux distribution that is provided with the RasPi starter kit or purchased separately. This 4 GB card should also be plugged into the SD card holder prior to power being applied to the RasPi.

CAUTION **Inserting or removing an SD card while the RasPi is powered on is never a good idea. Bad things can happen including data corruption or worse.**

Also, now would be a good time to skip down to the section that discusses how to load your own

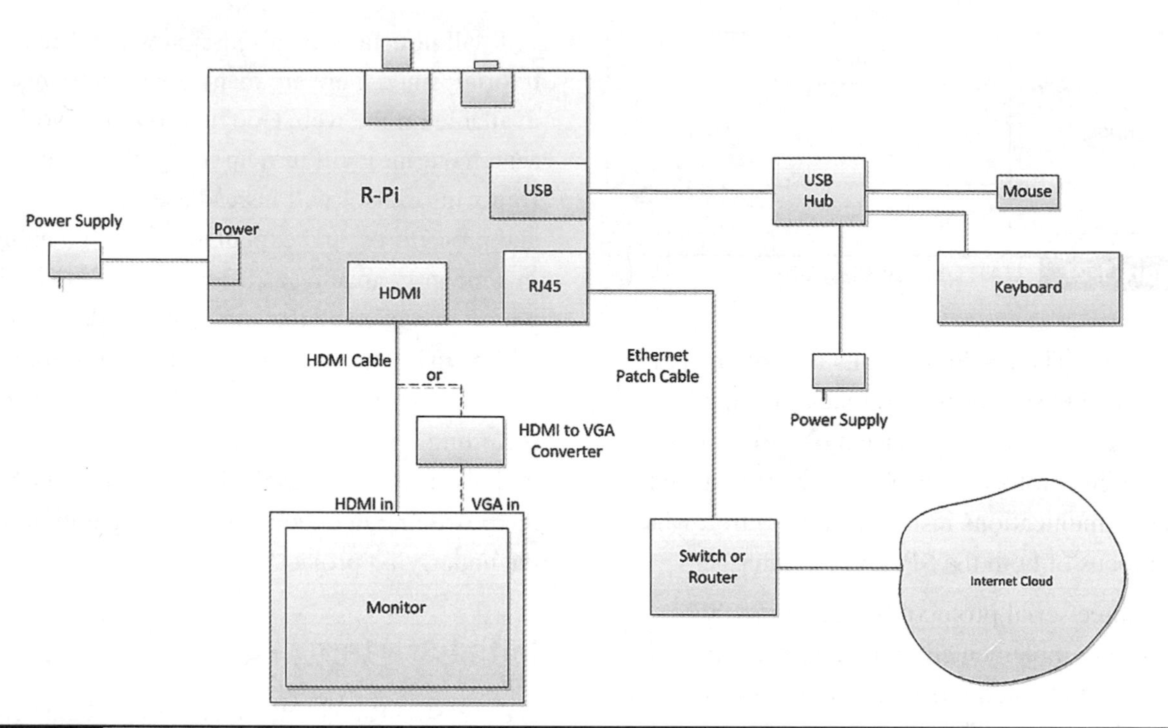

Figure 1-22 A suggested RasPi configuration.

Linux distribution on an SD card if you did not purchase a starter kit or a "prebuilt" SD card.

You should see the initial configuration screen, shown in Fig. 1-23, on the monitor after you connect the USB power to the RasPi. Please be patient; it takes a while. You will at first see a massive amount of text scrolling by on the screen, which will make absolutely no sense to you if you are not familiar with Linux.

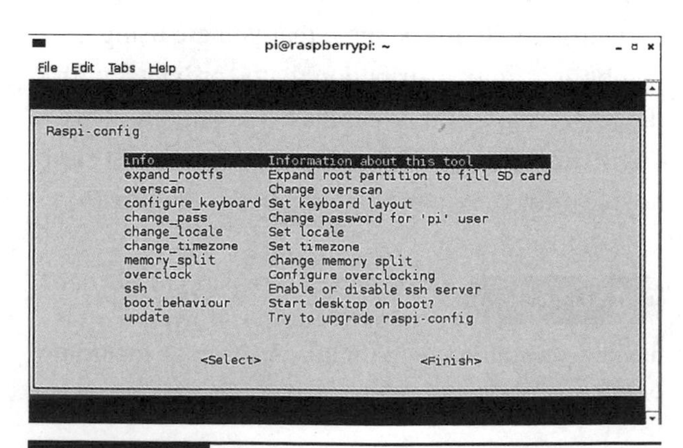

Figure 1-23 Raspberry Pi configuration screenshot.

Suggested configuration settings are shown in Table 1-8 along with some comments regarding why a particular setting was selected.

After you finish the configuration, the monitor prompt will eventually show. It is waiting for you to enter the user name, which is **`pi`**.

```
pi@raspberrypi~$ pi
```

Next, the password prompt shows **`Password:`**. Enter **`raspberry`**.

```
Password: raspberry
```

Please note I am not giving away any secrets, as every unmodified RasPi Linux distribution is created with this default password. You may have changed it in the configuration menu; if so, enter that password.

Next type **`startx`** and press Enter.

```
startx ↵
```

Table 1-8 Suggested Configuration Settings

raspi-config menu item	Setting	Comment
info	n/a	Info about the raspi-config app
expand_rootfs (expand root partition to fill SDS card)	yes or enter	The RasPi uses about 2 GB. This option will reclaim all remaining SD memory for use.
configure_keyboard	select your keyboard	Leave as default if your keyboard is not listed
change_pass	no	Leave the default password for now
change_locale	no	Default is "all locales" That's sufficient
change_timezone	select your time zone	
memory_split	no	Default of 192 MB for ARM and 64 MB for GPU is OK
overclock	no	For reasons mentioned earlier in chapter
ssh [enable or disable *secure shell* (SSH) server]	enable	Needed to run remote with PC and RasPi
boot_behaviour	yes or enter	Boot into the GUI desktop
update	yes or enter	Performs an update to the Linux distribution. This will take several minutes, and you must be connected to the Internet

This will create the GUI desktop, as shown in Fig. 1-24.

Congratulations, it is now about 15 pages into the book, and you now have the first indication that you have a working and useful Linux computer. Fear not; progress will be much faster from now on. To partially accomplish this goal, I will minimize the number of screenshots and simply use text to show you what to enter and how the computer responds.

Figure 1-24 Raspberry Pi GUI desktop.

Preparing your own SD card using a Windows PC

You will need two items other than the card itself. First you will need to download and install a program named win32diskimager.exe. This program is available at http://sourceforge.net/projects/win32diskimager/. The file is in a compressed Zip format from which you have to extract the program before running it. Note that in spite of the win32 in the name, this app works without a problem on my Win7, 64-bit laptop.

The second item you will need is the image file for the RasPi Linux distribution that you desire to install. The current version, at the time of this writing, may be downloaded from the main Raspberry Pi website http://downloads.raspberrypi.org/images/raspbian/2012-10-28-wheezy

-raspbian/2012-10-28-wheezy-raspbian.zip. It is a very large Zip file (647 MB) from which the Linux distribution must be extracted before it can be used as a disk image file. The Raspberry Pi Foundation currently appears to be updating the Wheezy-Raspian Linux distribution almost every month. This is subject to change, so take advantage of it while it lasts.

It is now a simple matter to create your own SD card once you have the image and the disk writer program. Insert a blank SD card into the PC, run the app, and then browse to the location where the image is stored. Then click on Write, and you will get a warning about destroying existing data on the card. Click OK and wait. It takes several minutes to create the image. You now have your own Linux distribution on an SD card. Also, remember that you can rewrite an SD card as often as you want, so feel free to experiment with different distributions.

I now need you to create a new SD card, both to ensure that you understand this process and that you have a specific distribution available that will also support the Chap. 10 book project, which requires what is known as the "1-Wire" serial protocol. This distribution is named Occidentalis V0.2 and is available at http://learn.adafruit.com/adafruit-raspberry-pi-educational-linux-distro/occidentalis-v0-dot-2. This distribution was created by the kind folks at Adafruit, where I purchase most of my RasPi goodies. The unusual name derives from the Latin name *Rubus Occidentalis* for the black raspberry, which is apparent from the GUI desktop that appears when this distribution is running, as shown in Fig. 1-25.

Figure 1-25 Occidentalis GUI desktop.

Some Linux Fundamentals

As I promised you earlier, I am not going to provide a tutorial on Linux in this book. However, you will still need to have some very basic knowledge of it to understand what is happening with the commands being processed. The discussion below is for readers with a very limited knowledge of Linux. Feel free to skip this section if you already have a basic to good understanding of Linux.

The Linux operating system is based upon Unix, and it has assigned built-in privileges, which limit most users to some extent but allow one user unlimited access. This unlimited user is named *root* and essentially is equivalent to an administrator level in a Windows operating system. Some commands can be run or executed only by root, again for security reasons. There is a fundamental computer security principle known as "least privilege" by which users are granted only as much access or privilege as they need to complete their task.

It is not considered a good idea to run all tasks or programs as root, so most of the time you will be running as the user named pi. However, as I mentioned earlier, some commands can only be run as root. The answer to this conundrum is to "promote" an ordinary user to what is known as a *super user*. The `sudo` command accomplishes this neat feat. I will frequently use the `sudo` command with other commands, and you now know why.

I will also typically use terminal windows to execute commands, as most of the time I will have the GUI desktop running. There are two flavors of terminal windows available, the *Lightweight X11 Desktop Environment* (LXDE) for *normal* users and a root level terminal. The only effective difference between the terminal windows is that I have to type `sudo` in the LXDE terminal while this is not required in the root terminal, since it already operates at that level.

You will also need to create some Python code to program the RasPi. I use a very simple but effective text editor named nano. All you need to run the editor is to open a terminal window and type `nano test_my_project.py` if you wanted to either create or open an existing file named test_my_project.py. The editor program has all the important commands listed at the bottom of the editor window. For example, to save the editor buffer, you have to press and hold the control key while simultaneously pressing the `"o"` key. This is shown as `^o` on the help screen.

A few of the very common Linux commands are shown in Table 1-9. It might be helpful to refer to this table when you are trying to understand why I entered a particular command as you progress through the projects.

Python

I have chosen to use Python to program the RasPi for several reasons. First, and most importantly, it is a simple programming language to use, and most beginners can start to use it immediately with little to no trouble. Second, despite being simple to use, it is a real programming language and shares most of the attributes that are common with other high-level, high-powered languages, such as C++ or Java.

You should visit the official Python website, http://www.python.org where you will find a wealth of information regarding Python. This site should be your primary resource to use to answer any of your questions regarding the language. There are also many good Python books available, including *Python Programming for the Absolute Beginner,* third edition by Michael Dawson and *Think Python* by Allen Downey. Another useful reference would be Simon Monk's *Programming the Raspberry Pi: Getting Started with Python*. Dr. Monk's book is concise, with a brief but thorough introduction to Python fundamentals plus the bonus of a few projects at the end of the book.

Python is classified, in computer science terminology, as a high-level language, which, roughly translated, means that Python users are thinking in abstract terms. The C language, in comparison, is a fairly low-level language in which programmers must contend with bits, bytes, and memory locations. The concept of abstractions and

Table 1-9 Commonly Used Linux Commands

Linux Command	Description
.	Refers to current directory
..	Refers to parent directory
cat *filename*	Print out the text in *filename*
cd	Change directory
cd ..	Change to parent directory
ls	List current directory contents
mkdir *directory*	Make (create) *directory*
pwd	Print out the name of the current directory
rm *filename*	Remove (delete) *filename*
rmdir *directory*	Remove (delete) *directory*

abstract data types will become clearer as we start developing Python programs.

The Python language design encourages the creation and use of modules that can be reused. New functions, modules, and classes can easily be added. Software reuse is a key goal that all programmers should try to adopt. It makes no sense to keep reinventing the wheel when proven solutions have already been developed and are available for reuse. We will see this reuse in action in the very first project program.

Python is also an interpreted language, which means that the code is executed line-by-line by the Python "engine" as it is encountered. Programs using interpreted languages typically run slower than programs that have been compiled and linked. This performance hit will not be an issue with any of the programs within this book. The advantage of interpreted programs is that they are considerably easier to modify and rerun as compared to compiled programs. All you have to do is make any needed changes in an editor, save the code, and then rerun the program in the Python shell. This significantly reduces program development time and, at the same time, increases your productivity and efficiency.

One other detail should be mentioned before I show you how to start developing your first Python program. The language is not named after the reptile but instead takes its name from the famous BBC show, "Monty Python's Flying Circus." Apparently, the Python creators were great fans of the show, and they currently encourage authors (or bloggers now) to incorporate related humour when writing about Python.

IDLE

IDLE is the name of an application that creates and runs the shell environment that I will use to develop and test your Python programs. Fig. 1-26 shows a portion of the desktop with two IDLE icons appearing.

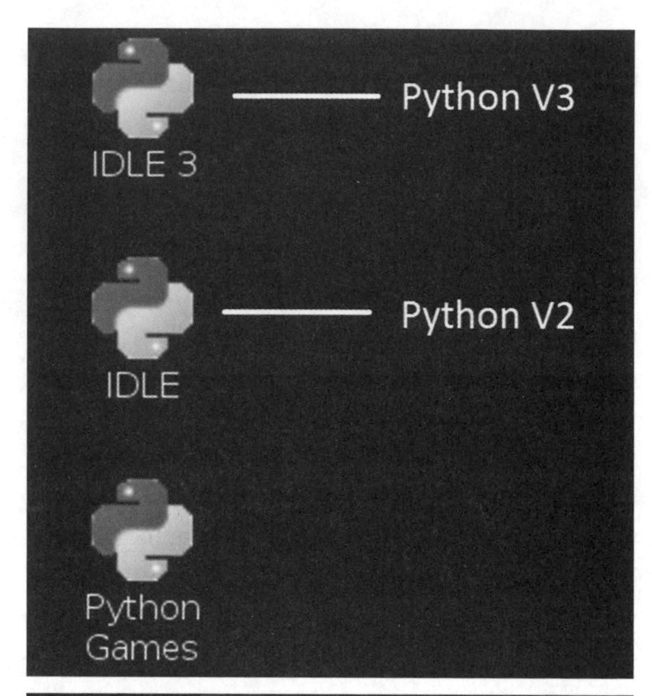

Figure 1-26 IDLE desktop icons.

The top icon opens a Python version 3 shell, while the icon directly underneath opens a Python version 2 shell. I will be using the version 2 shell, as that Python version is compatible with the software libraries that are needed to run the hardware used in the projects.

User interaction using the Python shell is intuitive; results are instantly displayed after an operation is performed and the Enter key is pressed. Adding 7 + 5 with the sum displayed below the input numbers is shown in Fig. 1-27.

Displaying text in the Linux shell is also easy to accomplish; simply use the print function. Traditionally, the first program to be run in most programming books is the so-called "Hello World" program. I do not like to trifle with tradition and will adhere to this unwritten rule. Now it is perfectly possible to execute the `print` command and see `Hello World` displayed below, as is shown in Fig. 1-28.

As this book is printed in monochrome, I will point out the following as you observe the output in the Python shell. The word `print` is reddish-

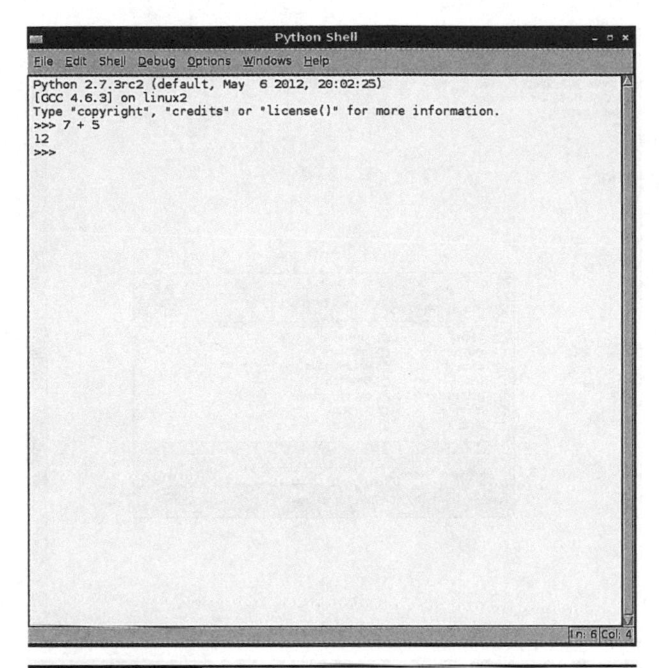

Figure 1-27 A simple Python operation.

orange, as it is a reserved word describing a preset function. The words `Hello World` in the parentheses following the `print` function are shown in green to indicate a string of characters. Character strings are enclosed between single quotes. Finally, the words `Hello World` displayed below the `print` function are in blue, as they represent a string output. This default color coding of program code is fairly standard with various development tools; however, the exact colors assigned to the different elements will vary.

From this point on, I will now use text only to show the shell prompt, commands, operations, and results to conserve valuable book space. You should also carefully observe the Linux shell because there may be information shown that I do not transcribe.

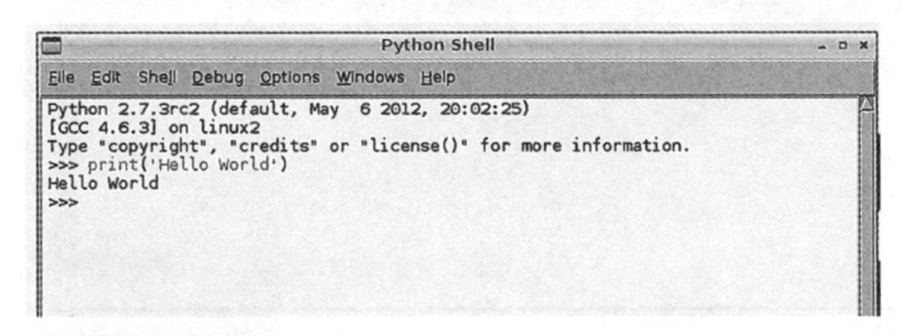

Figure 1-28 Python print command.

I will now show you how to create a very simple program that produces the same result as discussed above. I will use the nano editor to write the program and save it. I will then recall the saved program and run it from the Linux shell.

Open an LXDE terminal window and type:

```
nano
```

Next type:

```
print ('Hello World')
```

Press the key combination Ctrl and the letter `o`. (I will show this as `^o` from now on.)

```
^o
```

This action will bring up a prompt at the bottom of the editor, requesting the name of the file in which to save the buffer's contents. Type:

```
Hello.py
```

The .py is a standard Python program extension. Fig. 1-29 shows the nano editor at this point in time.

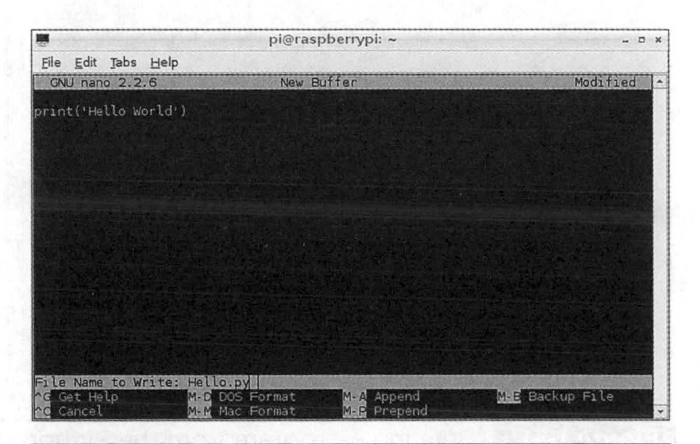

Figure 1-29 nano editor.

Type

```
^x
```

to exit nano. There will now be a Python program named Hello.py in your home pi directory.

A Python shell must now be opened to allow the file that I just created to be opened and run. Opening the file is done by clicking on the File menu tab and then clicking on Open from the drop-down menu. A file browser dialog box will be opened that shows the contents of the pi directory. Select the Hello.py file, as shown in Fig. 1-30.

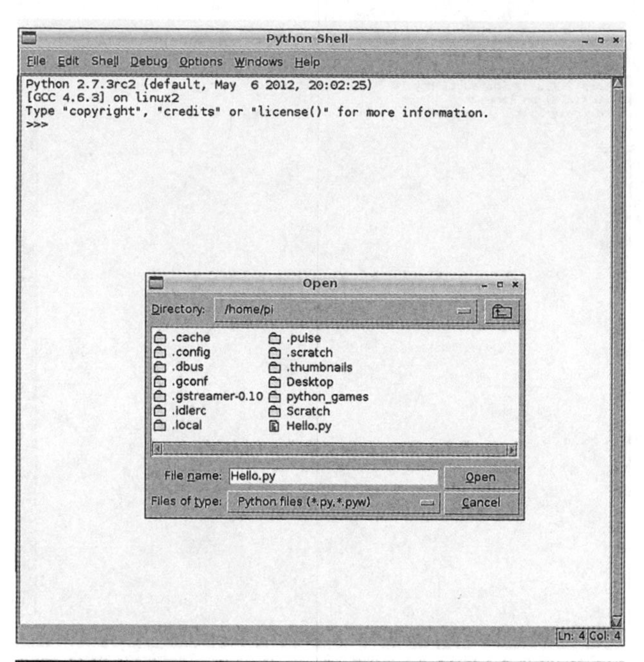

Figure 1-30 Opening a file in the Python shell.

Selecting the Hello.py file will cause a second window to appear on the desktop with the contents of Hello.py shown in the window. This is very convenient in that you can modify the file contents without affecting any of the content happening in the Python shell. To execute the program, you must be in the second window that was just opened where you can either open the Run menu tab and select Run or simply press the F5 function key. The Hello.py program results appear in the Python shell, as can be clearly seen in Fig. 1-31.

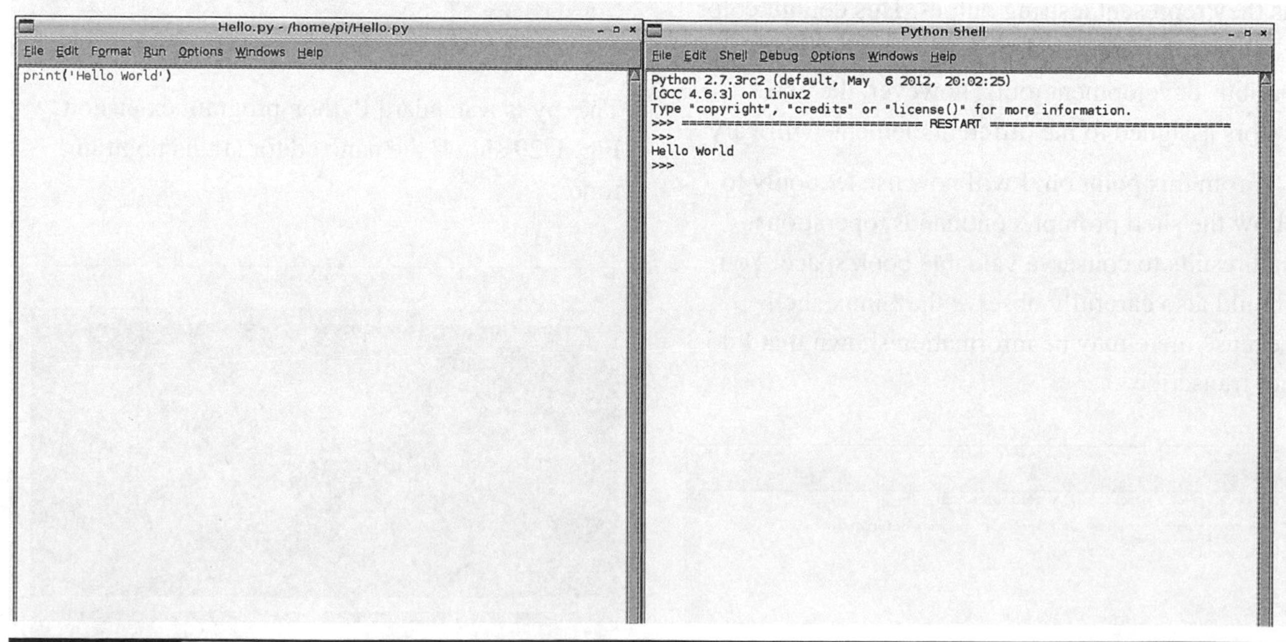

Figure 1-31 Running a program from the Python shell.

Summary

I have covered a lot of material in this chapter, from introducing the RasPi and a bit of its history to explaining the role that its inventors would like it to fulfill. I also covered the hardware aspects, as you need to understand the design decisions that went into the RasPi and the consequent constraints and attributes that you must consider when incorporating this board into a real world project.

A brief Linux introduction was covered to get you started in using this great operating system. Most people find that once they become proficient in using Linux, especially at the command-line level, they look at MS Windows with a newfound disdain. I am not pooh-poohing Windows; I am simply saying that Linux gives you much greater control of your programming environment than you could achieve by using Windows.

Next I discussed Python and demonstrated how simple it is to start programming the RasPi with the traditional "Hello World" program. Using the Python shell named IDLE just makes the whole effort very easy and, I hope, enjoyable.

CHAPTER 2

LED Blinker Project

Now we start creating projects using the RasPi. However, I still have some preparatory steps to discuss and information to provide about the prototype board that will be used in this and other projects. I will also present a detailed discussion of the GPIO library, which is critical to making all of the projects work. The first project will consist of turning on an LED connected to a GPIO pin and then causing that LED to blink.

Prototype Board

Let's focus on a piece of hardware that I will use to demonstrate how to blink an LED using the RasPi and some Python code. This section is entitled "Prototype Board" and that is exactly what I will be using. The board is called the Pi Plate and is available from Adafruit at https://www.adafruit.com/products/801. It comes as a kit of parts; however, a fully assembled Pi Plate mounted on a RasPi is shown in Fig. 2-1.

The Pi Plate attaches to the 26-pin GPIO connector, as you can see in the figure. All the active connections are brought out to the top of the Pi Plate, both to 0.1-inch, on-center, female header strips as well as to 3.5-mm screw clamp connectors. This makes for a very easy access to the GPIO pins. Please note that the pins on the Pi Plate are marked with both RasPi and BMC designations, which can be confusing. Please use Table 1-2, from Chap. 1, to cross-reference all the designations and ensure that you know what connection you are making.

CAUTION The Pi Plate does not provide any electrical buffers whatsoever, so be extremely careful and ensure that you do not accidently apply 5 V to any RasPi pin.

Soldering

I thought this would be an appropriate time to discuss soldering, since you will need to assemble the Pi Plate. The assembly operation is very straightforward, and Adafruit provides a step-by-step illustrated assembly guide at http://learn.adafruit.com/adafruit-prototyping-pi-plate. I have also included a photo of my trusty soldering station in Fig. 2-2.

It's not the prettiest or the snazziest station available, but it has served me well. The key to quality soldering work is to have good soldering technique, keep the soldering iron tip clean, and use the highest-quality solder available. Figure 2-3 shows the essence of good soldering technique. It is vital that the solder joint be hot enough for the solder to flow easily. It takes practice to apply just the right amount of solder; too little may result in a cold solder joint, and too much could lead to a short between closely spaced components.

Another issue regarding a good solder joint is the use of lead-free solder. Now, don't get down

Figure 2-1 Pi Plate mounted on Raspberry Pi.

on me; I am all about maintaining a healthful environment, but the elimination of lead from solder often produces poor solder joints unless some extra precautions are taken. The simplest and probably the best approach is to apply a high-quality, acid-free, solder flux to the joint prior to heating the joint with the iron. This will allow the lead-free solder to flow more freely and produce a better-soldered connection. Again, it takes practice to perfect soldering techniques.

One final thought that relates to solder joints as well as to other types of electrical connections is worth sharing. There is a long-running anecdotal observation that 90 percent of all electrical/electronic malfunctions are related to connection

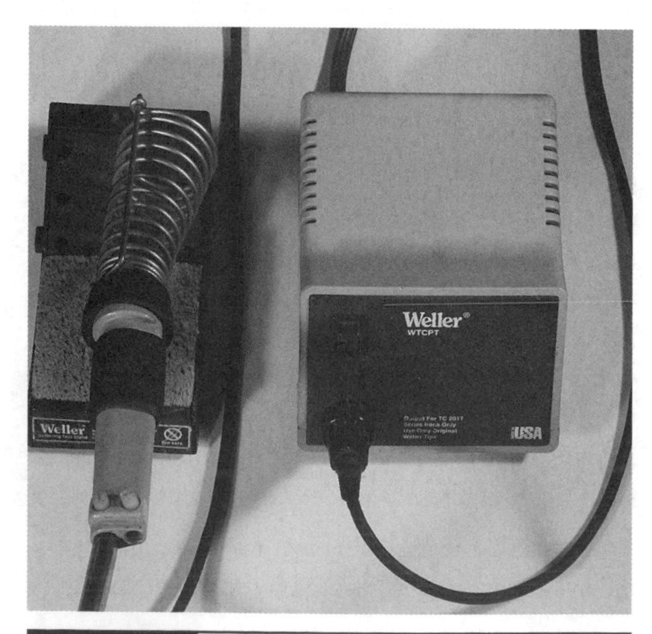

Figure 2-2 Trusty soldering station.

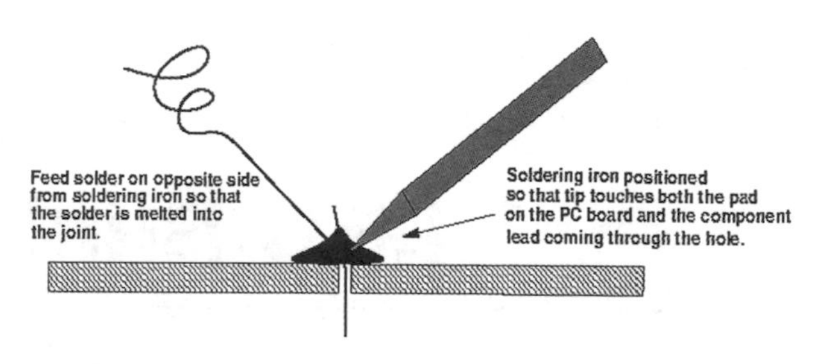

Figure 2-3 Good soldering technique.

malfunctions. This makes a lot of a sense when you think about it. We live in an oxygen-rich atmosphere and oxygen is a great reduction agent; it wants to oxidize every element it can possibly chemically combine with. Metal oxides are reasonably good insulators because some of their free electrons have been "taken" up by oxygen molecules. This leads to higher and higher resistance being built up in a connection that will eventually cause a failure.

Of course, current flowing through a resistance produces heat, which in turn can cause a fire if the current is sufficiently high. So what is the solution? One expensive solution is to gold-plate electrical contact surfaces. Gold does not oxidize and is not subject to this type of failure. It is, of course, very expensive and not practical for large-scale connectors. Another solution that the power industry employs is what is known as gas-tight connections, which require the use of special components and specialized training to produce. For the type of projects that I work on, I can only ensure that the solder joints are sound from both a mechanical and an electrical perspective. I also inspect electrical connections for oxidation and foreign matter and take appropriate action to replace the compromised connection or component.

Accessing the GPIO Pins

The GPIO pins have to be made programmable in order to blink the LED. Being made programmable means that the pins are set to either sense an input voltage or set an output voltage. The pins are accessed by using a high-level name such as pin (18) that you can readily identify—in this case, GPIO pin number 18. We must use a software library that contains the required high-level abstractions and hardware associations to enable us to control the GPIO pins with a Python program. This library is named Rpi.GPIO-0.4.1a.tar.gz and is readily available as a Linux archived file from http://code.google.com/p/raspberry-gpio-python/downloads/list. Figure 2-4 shows this website.

The archived file must be downloaded, extracted, and stored in an appropriate directory. I would suggest creating a new directory named Python located in the pi parent directory.

Open an LXDE terminal window and type the following commands in the terminal window:

```
pi@raspberrypi~$
pi@raspberrypi~$ pwd ↵
   —displays the current directory.
     /home/pi
pi@raspberrypi~$ ls ↵
   —list the contents of the current directory.
     Desktop python_games
pi@raspberrypi~$ mkdir Python ↵
   —make (create) a new directory named
     Python
pi@raspberrypi~$ ls
Desktop Python python_games
   —new directory now present
pi@raspberrypi~$ cd Python ↵
   —go into this new directory (change
     directory)
pi@raspberrypi~$ pwd ↵
   —check for the new path
/home/pi/Python
   —yes, it is correct
pi@raspberrypi~$ ls ↵
   —check if anything is in the new directory
pi@raspberrypi~$
   —prompt returns without displaying
     anything. directory is empty
```

You should now be in your development directory, Python, ready to download the GPIO library archive. You have several choices on how to download the archive file. You can use a browser on the RasPi and just do a direct download, or you can use a PC and download it into a thumb drive, which you then carry over to the RasPi and transfer it using the File Manager app. I choose to do the latter as the PC download process is much faster. In days of "yore" this approach was called the "sneakernet".

The archive must now be uncompressed and the contents extracted. Assuming you are still using a terminal window in the Python directory, type the following:

```
pi@raspberrypi~$ tar –xzf RPi.GPIO-
0.4.1a.tar.gz ↵
```

Ensure that you type in the exact name with uppercase and lowercase letters as shown. Linux is very picky in this regard and will throw an error message that it cannot find the file if you don't

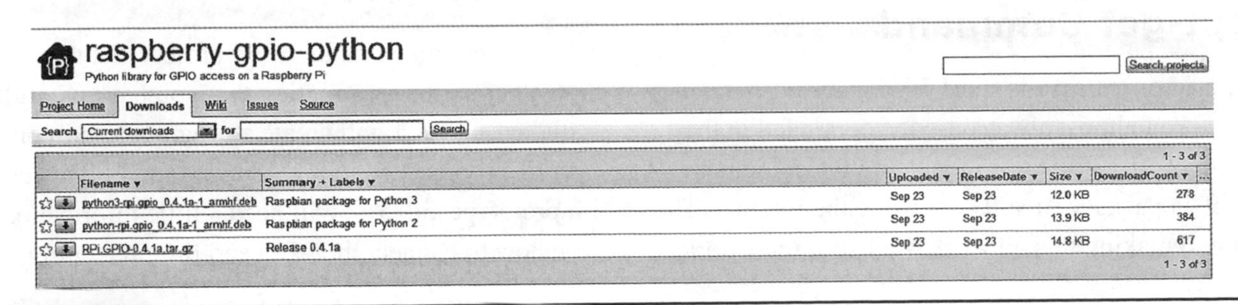

Figure 2-4 GPIO library download website.

enter the name exactly as it appears. The tar app will create a new directory named RPi.GPIO-0.4.1a, where you will find over a dozen files and directories newly created. Type the following to transfer into this directory.

```
pi@raspberrypi~$ cd RPi.GPIO-0.4.1a ↵
```

Incidentally, I will no longer explicitly show these commands, as you should now be more comfortable using them. I will instead simply say `cd` into `RPi.GPIO-0.4.1a`, and you will instantly know what I mean.

Let's test this newfound confidence! `ls` the `RPi.GPIO-0.4.1a` directory. You should now be looking at a list of all the files and subdirectories. One file should pique your interest; it is named INSTALL.txt. cat this file to read the contents. You could also double click on it to open it in a text editor, but I am feeling confident that you are starting to get the feel of how to operate quickly and efficiently at the command-line level, where the pros live. Back to the INSTALL.txt file. It contains important instructions regarding loading another module required to use the GPIO library. The key instruction is the following:

```
sudo apt-get install python-RPi.gpio
↵
```

Please read the following section before you enter the above command.

apt-get Command

The acronym `apt` is short for *advanced packaging tool*. You almost always use the command in the form of `apt-get` for instructing the computer to "get" a package of software using the Internet. The key to making this all work is a list of repositories that `apt` refers to in order to retrieve the requested package. The repository list is located at /etc/apt/sources.list in the Linux distribution.

The `apt-get` command also wants to know what you want done with the software package once it has been retrieved. That's the purpose of the `install` portion of the command. But `apt` has other neat features including the ability to update all the packages already installed in the computer. Simply type the following:

```
sudo apt-get update ↵
```

All available updates for installed packages will now be installed. Be patient; this will take a while, especially if there are many packages involved.

However, that's not all `apt` is capable of doing on a "global" scale. You can upgrade all your installed packages, provided upgrades are available. Type the following:

```
sudo apt-get upgrade ↵
```

Upgrades as the name implies are new versions of installed packages. You should always update prior to upgrading in order to lessen the chance that an inappropriate update will be applied to a newer version.

There is a great deal of information available on `apt` at www.debian.org/doc/user-manuals#apt-howto.

LED Project

I will now show you the LED project that ties together all the information that you have carefully studied so far . I will create a program to turn on an LED connected to pin 18 using the Pi Plate as a prototype aid. You will need a little information on how to connect the LED and limit the current flowing through the device. Figure 2-5 is a diagram showing the LED connections as well some physical descriptions that should be helpful to you in understanding the circuit.

The LED anode has a longer lead that is connected to the pin 18 screw terminal. The LED's shorter lead is the cathode, and it is connected to one lead of the 570-ohm (Ω) resistor. The other resistor lead is connected to the ground screw terminal. The resistor's value was calculated as follows:

- Current range that a RasPi pin can handle is 2 mA to 16 mA. So, 5 mA was selected as a low- to mid-range value with just enough current to operate the LED with a dim red light. No sense in overstressing the RasPi.
- The high output voltage from pin 18 is 3.3 V. The LED has a nominal drop of 0.7 V. Therefore, 3.3 – 0.7 or 2.6 V must be dropped with a 5 mA current.
- Ohm's law: $R = E/I$ where R stands for the resistance of a conductor in ohms, E stands for the potential difference across a conductor in volts, and I stands for the current through a conductor in amperes . Plugging in the values for E and I, we get 2.6 V/(5 mA ÷ 1000) = 520 Ω resistance. Note that the 5 mA had to be converted to amperes by dividing by 1000.
- 570 Ω is the nearest standard value resistor to 520 Ω.

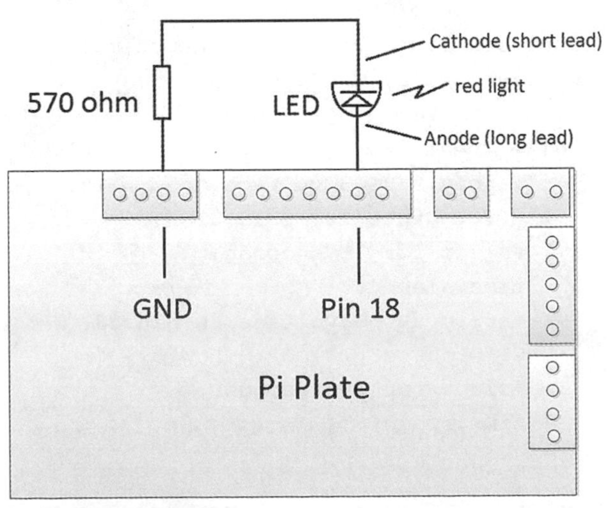

Figure 2-5 Diagram illustrating LED connection to the Pi Plate.

Figure 2-6 Actual LED connection.

Figure 2-6 shows the actual connections. I simply twisted the cathode lead to a resistor lead and soldered them to ensure a good connection.

Project Software

The program to turn on the LED is very straightforward. You should realize that when the RasPi is first powered on, all the GPIO pins are set to be in an input state, in other words, they will only sense or detect voltage not output a voltage. Therefore, we must change the state of the pin selected to be an output and then apply a high voltage to the pin. This is the fundamental logic behind the program.

Here is the program to be created using the nano editor. Comments are placed above each line to inform you what the following line is doing. Comments begin with the # symbol. All comments are ignored by the Python interpreter:

```
#import the library that allows us to
logically access the GPIO pins.
#refer to the library as GPIO. It is
like an alias.
import RPi.GPIO as GPIO
#use the BCM pin numbering scheme as
discussed in Chap. 1.
GPIO.setmode(GPIO.BCM)
#set pin 18 to be in the output mode.
GPIO.setup(18, GPIO.OUT)
#apply a HIGH voltage to pin 18
GPIO.output(18, GPIO.HIGH)
```

That's it, just three lines of code not counting the import statement that is really part of the program initialization. You should create the code in the nano editor and save it with a name that makes sense. I saved mine as Pin18_Test.py. Remember to use the .py suffix, since IDLE checks it to ensure that it is a valid Python program.

You need to open and run this program in the Python shell. However, there is an important action that you need to take or Python will complain and not run the program. You must run IDLE at the root level, not at a "normal" user level. This is needed, I believe, because the GPIO library is accessing Linux functions or resources that can only be accessed as root. Do the following to run the IDLE Linux shell as root:

Open a terminal window and type:

```
sudo idle ↵
```

That is all that's needed, and you will be at the exact place where you can open and run your program, as discussed in a previous section. Running the program will turn on the LED; not very exciting so far, as our journey begins with small steps. The next section shows you how to blink the LED.

Blinking the LED

A variety of approaches can be taken to blink the LED. The approach I took was simple and direct. The logic was to turn on the LED; wait a set amount of time; turn off the LED; wait a set amount of time; and then continually repeat this process forever.

Waiting a set amount of time may be accomplished in several ways. The way I selected was to use a function named `sleep(seconds)` that is part of the time library. All you need do is provide the amount of time, in seconds, that you want the RasPi to do nothing—essentially spinning its wheels. The `seconds` argument, the value put in the parenthesis, is a real number, i.e., one with a decimal point. I used .5 second for my initial program. The modified program, with comments, is shown below.

```
#import the library that allows us to
logically access the GPIO pins.
#refer to the library as GPIO. It is
like an alias.
import RPi.GPIO as GPIO
#import the time library to access the
sleep function.
#use the BCM pin numbering scheme as
discussed earlier in the chapter.
GPIO.setmode(GPIO.BCM)
#set pin 18 to be in the output mode.
GPIO.setup(18, GPIO.OUT)
#loop forever.
while True:
    #apply a HIGH voltage to pin 18.
    GPIO.output(18, GPIO.HIGH)
    #sleep for .5 seconds.
    time.sleep(.5)
    #apply a LOW voltage to pin 18.
    GPIO.output(18, GPIO.LOW)
    #sleep for .5 seconds.
    time.sleep(.5)
```

While you can make the modifications to the original program using the nano editor, I found it more convenient to use the shell editor, which displays the program when you first open it. This editor will allow you to save the program with the original or a new name.

There are a few items in the program that should be explained. The time library is accessed by using the `import` statement, the same way the GPIO library was accessed.

The looping behavior is imposed by two things; First, I used what is known as a conditional statement, `while True`. The `while` portion of this statement checks what immediately follows it, and if the expression evaluates to a logical `True`, will loop or repeat all statements below the conditional that are indented at least four spaces. The logical value of `True` that follows the `while` statement thus causes the looping to continue forever.

Pausing the program is a result of calling or invoking the `sleep` function, a part of the time library. The actual call is the statement `time.sleep(.5)`. This is not the first time you have seen this call type; I used it several times in the original program, i.e., `GPIO.setup(18, GPIO.OUT)`. The period in the statements represents a special operator known as the "member of" operator. Restating the `time.sleep(.5)` call with this operator translates to "call the sleep function that is a member of the time library with an argument of .5." Don't worry if this discussion is somewhat confusing; all I am trying to do is to gently introduce you to the concept of object-oriented programming, something that I will pursue in later chapters.

Now back to the Blinker program. Make the modifications to the original program and then run it. You should be rewarded with a blinking LED for all your efforts. Congratulations! Show all your family and friends!

Summary

I started this chapter with a discussion on the Pi Plate prototype plate that serves as an excellent project experimental platform. Next, I briefly mentioned some good soldering tips that have helped me over the years. Next, there was the discussion regarding the GPIO library that may have overwhelmed some of you a bit, but that was necessary in order to set the stage for the LED project. The good news is that we will simply use the GPIO library from this point on in various projects. That's software reuse at its best. The last sections of this chapter focused on a real world project of first turning on an LED and then blinking that same LED.

Now we move on to bigger, better, and more interesting projects. And yes, you will still learn and gather a good education as you progress through the remaining projects.

CHAPTER 3

MP3 Player Project

In this project, I will show you how to create a fairly simple, yet fully functional, MP3 audio player. I will also describe how to use the GPIO pins as inputs, which will allow the player to use some hardware push-button switches to select the MP3 songs it will play.

We will also explore the *Advanced Linux Sound Architecture* (ALSA) software package that drives the Linux sound production and provides some utilities essential to helping us complete this project.

Prototype Connector

Let's first focus on a piece of hardware that I will use to connect the project hardware switches with the RasPi's GPIO pins. This prototype aid is different from the Pi Plate introduced in the last chapter. Called the Pi Cobbler, it is shown in Fig. 3-1 without the interface ribbon connector in order to show all the pins clearly.

Really just a direct extension of, or a breakout kit for, the GPIO connector, it allows all the pins to be connected to a solderless breadboard. It is available as a kit from a variety of sources, including Adafruit at www.adafruit.com/products/914. A fully assembled Pi Cobbler mounted on a breadboard and connected to a RasPi is shown in Fig. 3-2.

Using the Pi Cobbler along with the breadboard will allow for rapid project construction and easy modifications to the existing project. It will help you to have a variety of prestripped wire available to use with the breadboard. You can either make your own using 22-gauge, solid pickup wire or purchase a package from your favorite electronics components supplier. Figure 3-3 shows a package that I bought for the book projects. These wires are actually stranded for flexibility and are also terminated with a solid, insulated pin connection. It is well worth the cost to buy a kit of these wires.

Portable Stereo Speaker

You will need a way to listen to the MP3 recordings. I use the very small and inexpensive stereo speaker device, the Veho360 (www.veho-uk .com) that is shown in Fig. 3-4.

This device has a 3.5-mm jack that you can plug into the RasPi audio-out connector. It also has a rechargeable battery that can be recharged by plugging it into any standard USB socket. However, I would not recommend using the RasPi USB sockets due to the current limitations that I discussed in Chap. 1. The volume control on the speaker will allow you to adjust the volume in only two steps. All in all, this powered speaker is fairly decent in reproducing audio; however, I would not get rid of my home theater system for it.

ALSA

Advanced Linux Sound Architecture (ALSA) is the principal sound driver that is used in most of the

Figure 3-1 Pi Cobbler prototype connector.

RasPi Linux distributions. It is very stable software that has been widely adopted for use in various Linux distributions. It is totally open source and is strongly supported at www.alsa-project.org. The eight packages that currently make up ALSA and a brief descriptions of each are listed in Table 3-1.

Although the ALSA drivers should be preinstalled with the distribution, you should execute the following command in a terminal window to ensure that the drivers are in place:

```
sudo modprobe snd-bcm2835 ↵
```

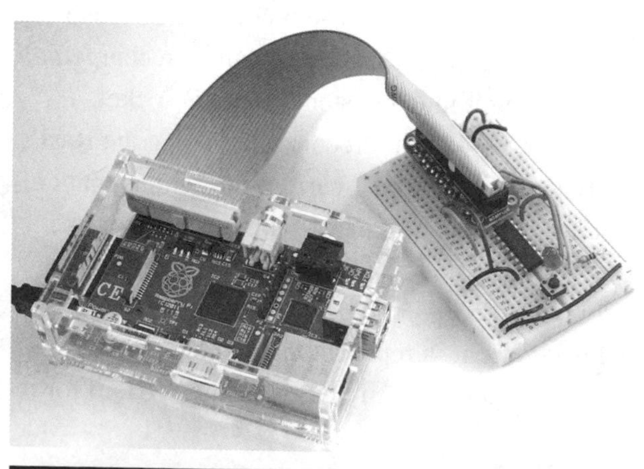

Figure 3-2 Pi Cobbler on a solderless breadboard connected to a Raspberry Pi.

Figure 3-3 High-quality breadboard interconnect wires.

The app modprobe loads *Loadable Kernel Modules* (LKM), which, in this case, is the ALSA compliant sound driver, snd-bcm2835. You should also check a file named *modules* located in the /etc directory for the presence of a snd-bcm2835 item. This file is checked at boot time for all the LKMs that are to be loaded.

You also need to install the alsa-utils package as that contains several apps that are important to get our project working. The installation uses the `apt-get` command that was discussed in Chap. 1. Type the following command in a terminal window:

```
sudo apt-get install alsa-utils ↵
```

Figure 3-4 Veho360 speaker.

Table 3-1 ALSA Packages

Package Name	Description
alsa-driver	Low-level drivers connecting hardware with the Linux kernel
alsa-lib	A C language API for the drivers
alsa-utils	A collection of very useful utilities
alsa-tools	An advanced toolkit used with certain sound cards (not the RasPi)
alsa-firmware	Firmware to be installed on certain sound cards (not the RasPi)
alsa-plugins	Used with audio libraries and sound servers
alsa-oss	Open Sound System, an older version provided for compatibility purposes
pyalsa	An ALSA Python module used with a video-editing system

One more step is required to ensure that sound is produced at the analog audio jack. Type the following in a terminal window:

```
sudo amixer cset numid=3 1 ↵
```

The app *amixer* is part of the alsa-utils package that allows us to select the desired audio output connector. The HDMI is the default audio output device when the RasPi is first booted. The audio output is redirected using the above command. The number at the end of the above command line represents the following:

0—auto

1—analog

2—HDMI

Testing the Analog Audio

Now, it is time to test the audio from the RasPi. Connect a powered speaker to the 3.5-mm jack and type the following in a terminal window:

```
sudo speaker-test ↵
```

You should now be hearing a *rushing* noise from the speaker. This is a *pink-noise* signal being created by the speaker-test app that is part of the ALSA package. (For a more detailed explanation of pink noise, see the following section on Analog Audio Frequency Response.) If you do not hear this noise, review the commands listed above and ensure that they were entered as shown. The speaker-test app has many more options than just producing a noise output. Go to http://manpages.ubuntu.com/manpages/natty/man1/speaker-test.1.html for more information.

Analog Audio Frequency Response

Having a pink noise output provides an unexpected opportunity to measure the frequency response of the RasPi analog audio system. I have had many years of experience as an acoustical engineer, and to me, this seemed a natural opportunity to assess the acoustic performance with this particular subsystem. First, let us delve into a bit of background information regarding pink noise and why it is so useful in determining a system's frequency response. Noise, by its intrinsic nature is random, with energy spread uniformly across the observed frequency spectrum or bandwidth. This type of noise is also called *white noise* to point out the uniformity of the energy.

Frequency analysis for acoustical systems typically uses a set of filters known as one-third octaves that simulate to some extent the response of the human ear. This means that the filters are narrower at lower frequencies and become wider as the frequency increases. Applying white noise to this filter bank would result in an upward or positive

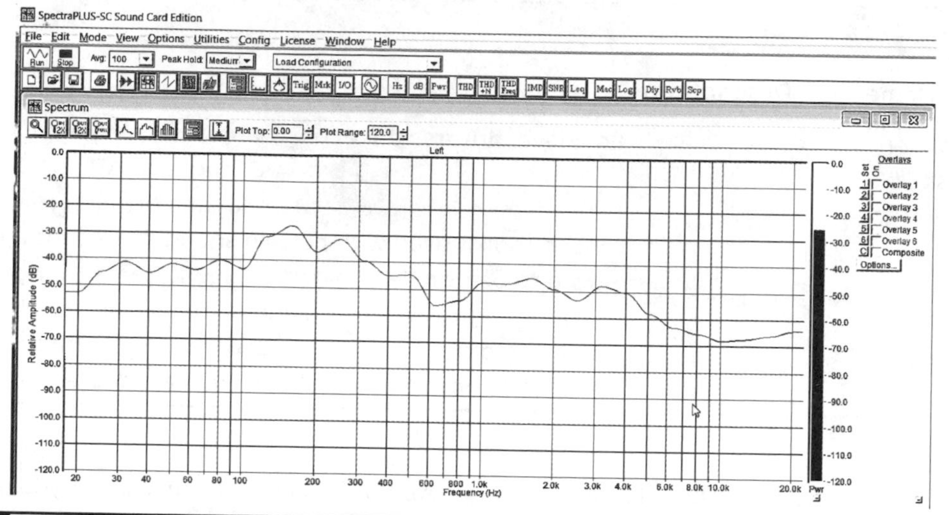

Figure 3-5 Raspberry Pi one-third-octave acoustic analysis.

sloped response. White noise is prefiltered (or predistorted if you may) such that the conditioned noise signal that comes out of the one-third-octave filters is a flat response. Any deviation from a flat response would, therefore, be caused by the system under test, not the applied noise signal. This preconditioned noise is called *pink noise*.

Figure 3-5 is a screenshot of an acoustical analysis program measuring the analog audio output producing pink noise from the speaker-test app.

The audio output is far from being flat as you can see from the irregular curve. The good news is the magnitude of these curves is not so severe as to appreciably distort the audio. Audio purists would probably use a multiband equalizer to compensate for these response deviations.

Audio File Formats

A brief discussion of the audio file formats will help you understand the pros and cons of using different formats when creating audio apps. There are three broad categories for audio file formats:

1. Uncompressed—An uncompressed storage uses the most storage of any format.
2. Lossless compression—Skips silent portions; stores audio content.
3. Lossy compression—Compresses everything; introduces a little distortion.

Table 3-2 lists some representative formats from each of these categories

WAV versus MP3 Formats

I will further discuss these formats, since the project uses the MP3 format. However, I have also included a simple demonstration of the RasPi playing a *Waveform Audio File Format* (WAV) file. The WAV format provides the truest reproduction of the original audio content as is possible using digital technology. The source content is sampled at a high rate using a sufficiently large number to precisely record the amplitude of the digital sample. High sample rates and large data-storage numbers mean very large storage requirements. For instance, a three-minute symphonic musical score might take 30 MBs of storage. In comparison, an MP3 format of the same content might take about 2 MBs, a 15:1 compression ratio. Of course, the trade-off is the quality of the music reproduction. Most people find the MP3 format acceptable, especially if reproduced on a device, such as a smartphone or tablet computer.

Table 3-2 Audio File Formats

Name	Compression Type	Compression Ratio	Remarks
WAV	Uncompressed	1:1	Also called PCM
AIFF	Uncompressed	1:1	
AU	Uncompressed	1:1	
FLAC	Lossless compression	1:1	
WMA Lossless	Lossless compression	2:1	Windows Media Audio
M4A	Lossless compression	2:1	Apple lossless format
MP3	Lossy compression	10:1 to 20:1	
Vorbis	Lossy compression	10:1 to 20:1	
ACC	Lossy compression	10:1 to 20:1	
WMA Lossy	Lossy compression	10:1 to 20:1	

Furthermore, the RasPi audio reproduction is not the finest as previously discussed.

Playing a WAV File

It is quite easy to play a WAV file if you use another app contained in the alsa-utils package. This app is named *aplay*, and all you need do is call the app along with the name of the WAV file to be played. I have placed public domain WAV and MP3 files named class.wav and class.mp3 onto this book's website, www.mhprofessional.com/raspi. Please download them into the pi directory. You will use class.mp3 in the next section.

Now, open a terminal window and type:

```
sudo aplay class.wav ↵
```

Ensure that your powered speaker is attached to the analog audio jack and that the jack is selected per the `amixer` command line shown on page 33. You should now be hearing the beautiful sounds of a Bach fugue. The musical score lasts for about 2.5 minutes. It can be stopped at any time by pressing `^c`.

The MP3 version of the same music file is approximately 1.2 MB as compared to the WAV file that is over 25 MB in size. You will have an opportunity to play the MP3 version in the next section after I introduce you to a software MP3 player. That should give you an excellent opportunity to compare the sound quality of the uncompressed WAV versus the lossy compressed MP3.

Playing an MP3 File

There are many Linux-compatible apps available to play MP3 files. I have selected a file named mpg123, as it is very stable and is easily controlled from a command line, which is what we need for our RasPi MP3 player. Use `apt-get` to load this app as follows:

```
sudo apt-get install mpg123 ↵
```

NOTE **There is a clone of mpg123 available on the web named mpg321. I am sure that it works just fine; however, I prefer to stay with the original.**

Now type this to play the MP3 file that you downloaded:

```
sudo mpg123 class.mp3 ↵
```

The MP3 will play for precisely the same amount of time as the WAV version and can be stopped at any time by pressing `^c`, as was the case for the aplay app.

I could not detect any appreciable difference between the two formats, but you may; and it might also be fun to invite friends and family to see if they can detect differences between the audio file formats.

mpg123 is also known as a console player app, since you interact with it by using the keyboard and observing the app output on a terminal window. This is exactly what we need to create a RasPi-controlled MP3 player. It is important to see what terminal controls are available and what their respective functions are. This is easily accomplished by pressing the h key while the app is running. Figure 3-6 shows the result of pressing this key while an MP3 was playing.

Table 3-3 is an easy-to-read list of these terminal control keys with additional explanations.

Incidentally, it is very easy to play all the MP3s in a directory by typing the following:

```
sudo –vC *.mp3 ↵
```

The v in the command line will provide a verbose output and the C puts the app into a full control mode where all the terminal control keys are made operational. The * in the command line is a wild card commanding the app to play all the MP3s it can find in the current directory. When all is said and done, this is a very powerful but fairly simple MP3 player app that will suit our purposes quite well. All we need now are some switches to select the songs and some Python code to automate it all.

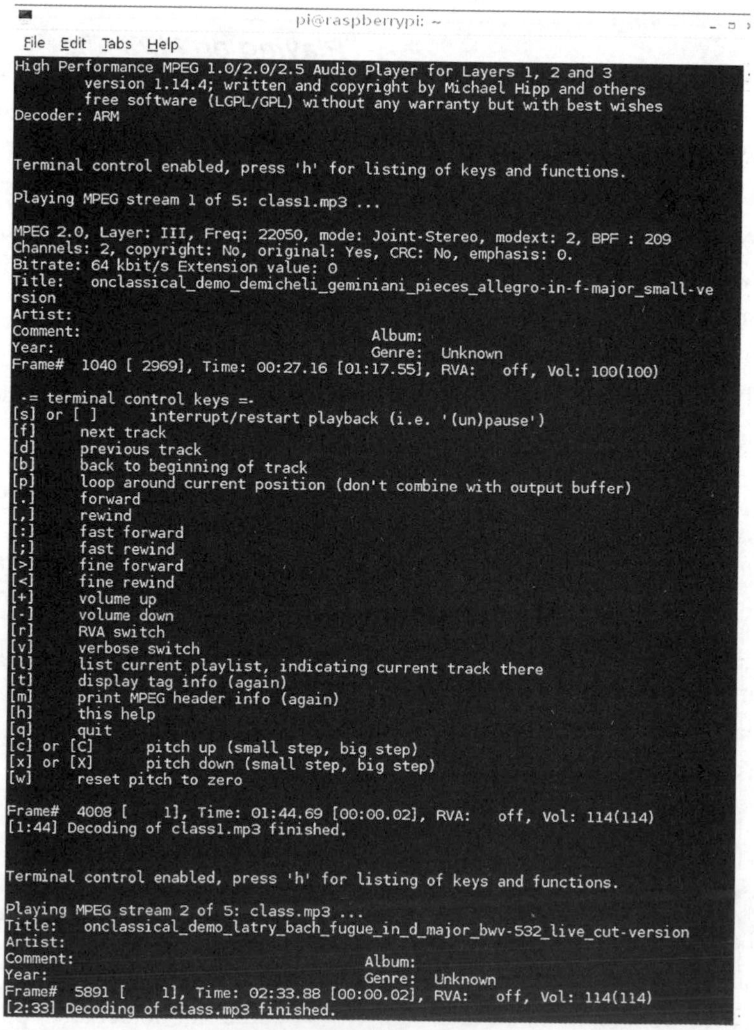

```
pi@raspberrypi: ~
File Edit Tabs Help
High Performance MPEG 1.0/2.0/2.5 Audio Player for Layers 1, 2 and 3
        version 1.14.4; written and copyright by Michael Hipp and others
        free software (LGPL/GPL) without any warranty but with best wishes
Decoder: ARM


Terminal control enabled, press 'h' for listing of keys and functions.

Playing MPEG stream 1 of 5: class1.mp3 ...

MPEG 2.0, Layer: III, Freq: 22050, mode: Joint-Stereo, modext: 2, BPF : 209
Channels: 2, copyright: No, original: Yes, CRC: No, emphasis: 0.
Bitrate: 64 kbit/s Extension value: 0
Title:   onclassical_demo_demicheli_geminiani_pieces_allegro-in-f-major_small-ve
rsion
Artist:
Comment:                                  Album:
Year:                                     Genre:  Unknown
Frame#  1040 [ 2969], Time: 00:27.16 [01:17.55], RVA:   off, Vol: 100(100)

 -= terminal control keys =-
[s] or [ ]      interrupt/restart playback (i.e. '(un)pause')
[f]      next track
[d]      previous track
[b]      back to beginning of track
[p]      loop around current position (don't combine with output buffer)
[.]      forward
[,]      rewind
[:]      fast forward
[;]      fast rewind
[>]      fine forward
[<]      fine rewind
[+]      volume up
[-]      volume down
[r]      RVA switch
[v]      verbose switch
[l]      list current playlist, indicating current track there
[t]      display tag info (again)
[m]      print MPEG header info (again)
[h]      this help
[q]      quit
[c] or [C]      pitch up (small step, big step)
[x] or [X]      pitch down (small step, big step)
[w]      reset pitch to zero

Frame#  4008 [    1], Time: 01:44.69 [00:00.02], RVA:   off, Vol: 114(114)
[1:44] Decoding of class1.mp3 finished.


Terminal control enabled, press 'h' for listing of keys and functions.

Playing MPEG stream 2 of 5: class.mp3 ...
Title:   onclassical_demo_latry_bach_fugue_in_d_major_bwv-532_live_cut-version
Artist:
Comment:                                  Album:
Year:                                     Genre:  Unknown
Frame#  5891 [    1], Time: 02:33.88 [00:00.02], RVA:   off, Vol: 114(114)
[2:33] Decoding of class.mp3 finished.
```

Figure 3-6 mpg123 terminal control key list.

Table 3-3 mpg123 Terminal Control Keys

Terminal Control Key	Action
s	Pause toggle (press once to pause, press again to resume)
f	Next track
d	Previous track
b	Back to beginning of track
p	Loop around current position
.	Forward
,	Rewind
:	Fast forward
;	Fast rewind
+	Volume up
-	Volume down
r	*Relative volume adjustment* (RVA) switch—uses the MPEG ID3V2 tag to adjust album loudness
v	Verbose switch
l	List current playlist
t	Display tag information
m	Print MPEG header information
h	Help list
q	Quit
c	Small pitch up change
C	Large pitch up change
x	Small pitch down change
X	Large pitch down change
w	Reset pitch to zero

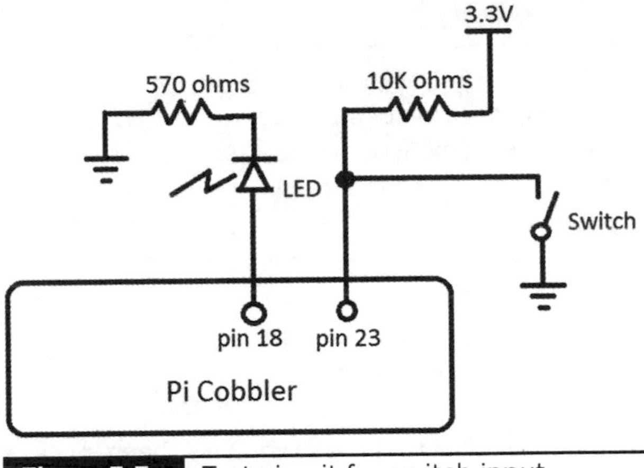

Figure 3-7 Test circuit for switch input.

Hardware Switch Inputs

Let's take a break from the all the software discussions and focus on how to connect hardware switches to the RasPi and how to respond when a user depresses a switch. Recall from Chap. 2 that all GPIO connectors can be set into either an input or output mode. For the LED blinker, I set a pin 18 as an output. In this section, I will set pin 12 as an input and leave 18 as an output. The goal is to have the LED connected to the pin 18 light when a switch connected to pin 12 is pressed. I will be using the 3.3-V power supply to trigger pin 12; however, it is worth repeating the warning made in Chap. 1.

Caution **All GPIO voltage levels are 3.3 V and are not 5 V tolerant. There is no overvoltage protection incorporated on the RasPi; and if you inadvertently subject a GPIO pin to 5 V, you will wreck your RasPi.**

There is 5 V present on the Pi Cobbler, and it is not too difficult to inadvertently connect to that power supply instead of the 3.3-V power supply. Just be extra careful and recheck your connections twice before powering up.

Use the schematic in Fig. 3-7 as a guide to connecting the LED, the switch, and the associated resistors that make up the test circuit. Use a solderless breadboard, as it readily accommodates the Pi Cobbler and other components.

Here is the Python code that will run this little circuit. This program is available on the book's website as Test_Switch.py.

NOTE **There are two equal signs required to do a comparison, i.e., `GPIO.input(23) == False`. If you make a mistake and put only one, it will become an assignment statement, which may or may not be evaluated to a logical state. An erroneous statement, `GPIO.input(23) = False`, would likely evaluate to False, since you cannot programmatically set an input. In any case, the program will run, but not operate as expected. This is called a logical error and is a common occurrence with beginning programming students.**

Test_Switch.py

```
#import the GPIO library to access the pins.
import RPi.GPIO as GPIO
#use the BCM pin numbers.
GPIO.setmode(GPIO.BCM)
#set pin 18 as an output.
GPIO.setup(18, GPIO.OUT)
#set pin 23 as an input
GPIO.setup(23, GPIO.IN)
#start a 'forever' loop.
while True:
    #check if the switch has been pressed.
    if(GPIO.input(23) == False):
        #set the pin to a HIGH output
        GPIO.output(18, GPIO.HIGH)
    #Not pressed, then set pin to a LOW output.
    else:
        GPIO.output(18, GPIO.LOW)
```

I have used an `if/else` conditional statement in this program that checks if pin 23 is LOW or False, and if so, will set pin 18 to GPIO.HIGH. Otherwise, the conditional statement will go directly to the `else` portion where pin 18 is set to GPIO.LOW.

Run the program and observe that each time you press the switch the LED light turns on and continues to stay lit as long as the switch is pressed. Figure 3-8 shows the actual circuit on the breadboard with the RasPi connected with the flat ribbon cable to the Pi Cobbler.

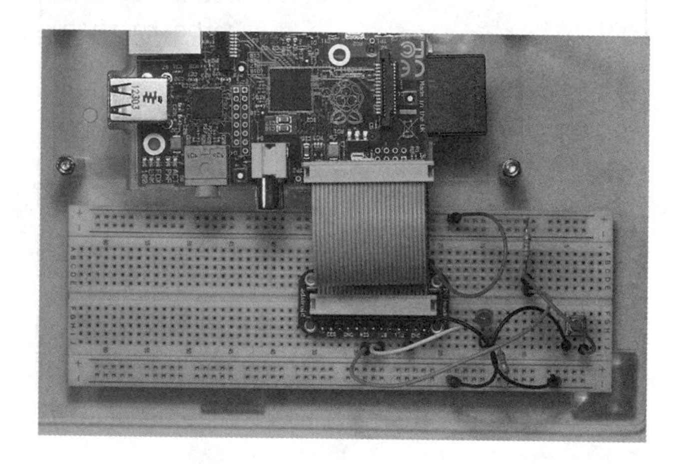

Figure 3-8 Actual switch test setup.

Thinking in RasPi Time

You must always think about the operational speed of the RasPi when creating a new program. The `while` loop in this program repeats over a million times a second, meaning that all the commands in the loop are constantly being repeated. That's the reason why I used an `if/else` conditional test statement to repeatedly set the pin value HIGH so that it matches the duration of the key press. If I had neglected to add the `else` portion, the first time the button was pressed, it would have lit the LED, but it would have remained on. Try deleting the `else` portion to confirm this behavior. But why does it stay on?

The answer is simple: without the `else` portion, there is no command to turn it off. The pin becomes latched so that it remains in the last state it was commanded to be in. That's what I mean by thinking in RasPi time. I have found that it is possible for new (and sometimes, not-so-new) microprocessor programmers to become very confused regarding this real-time aspect of programming. Controlling actual hardware can be a challenge. I will attempt to point out these *challenging* program areas in the book projects to help avoid confusion on your part.

MP3 Project Requirements

The first step in any project is to specify the requirements that must be met to satisfy the project user. In this case, the project requirements are rather simple. They are listed below:

- Use the mpg123 app as the MP3 player
- Play selected MP3 files (tracks) stored in a preset directory
- Use the computer monitor to display file (track) data
- Use the 3.5-mm analog audio jack for output
- Use the keyboard to enter commands to control the mpg123 app while it is playing a song
- Use one of three push buttons to play a preselected song

It is a straightforward process to build this project, as most of the component parts and software have already been discussed.

Writing the Program

The key issue is to start the mpg123 app from a Python program with all the needed arguments. Fortunately, Python provides a function named `system()` that allows us to do this quite easily. And, it is part of the OS package. The following Python program I created is named MP3_Player.py, and it is available for download from the book's website.

MP3_Player.py

```
#import the os package as it contains the system() function call.
import os
#import the GPIO library to access the pins.
import Rpi.GPIO as GPIO
#use the BCM pin numbers.
GPIO.setmode(GPIO.BCM)
#set pin 23 as an input.
GPIO.setup(23, GPIO.IN)
#set pin 24 as an input.
GPIO.setup(24, GPIO.IN)
#set pin 25 as an input.
GPIO.setup(25, GPIO.IN)
#ensure that the analog output is selected.
os.system('amixer cset numid=3 1')
while True:
    #check if the button connected to pin 23 has been pressed.
    if(GPIO.input(23) == False):
        #start the mpg123 app with full keyboard control and  screen output playing
        class.mp3.
        os.system('mpg123 -vC class.mp3')
    #check if the button connected to pin 24 has been pressed.
```

```
    if(GPIO.input(24) == False):
        #start the mpg123 app with full keyboard control and verbose output playing
        class1.mp3.
        os.system('mpg123 -vC class1.mp3')
    #check if the button connected to pin 25 has been pressed.
    if(GPIO.input(25) == False):
        #start the mpg123 app with full keyboard control and verbose output playing
        Maid.mp3.
        os.system('mpg123 -vC Maid.mp3')
```

MP3 Player Schematic

All the pin and switch connections are shown in Fig. 3-9. You can simply add on the two additional push-button switches and resistors to the existing circuit that was built according to Fig. 3-7. You can also leave the LED connected, if you so desire, as it will not be incorporated into this circuit.

Figure 3-10 shows the actual circuit on the breadboard with the RasPi connected via the flat ribbon cable to the Pi Cobbler and the Veho360 connected to the analog output jack. The three push buttons that select one of three MP3 songs are placed on the right side of the solderless breadboard.

Figure 3-10 Actual MP3 player.

Testing the Project

Load the MP3_Player.py program into the pi directory. Then open a terminal window and type:

```
sudo python MP3_Player.py ↵
```

I found that trying to use a Python IDLE shell to run this program might cause unreliable operation because the mpg123 app requires a direct keyboard input that the shell may interrupt. All the keyboard commands shown in Table 3-3 are available for use once the program is started. You should experiment with the various commands to gain an appreciation of the depth of functionality that the mpg123 app possesses.

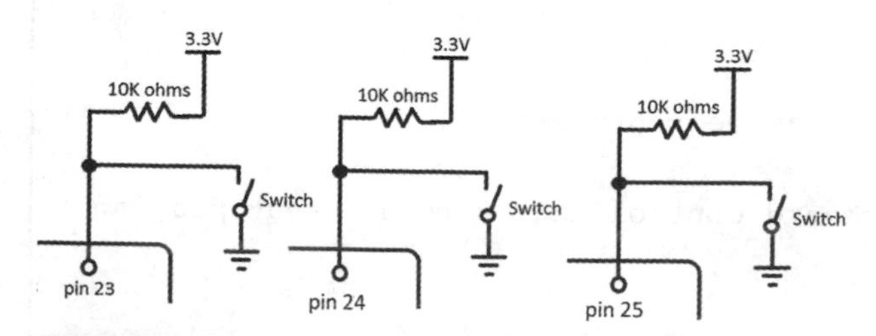

Figure 3-9 MP3 player circuit.

Figure 3-11 is a screenshot showing an MP3 player in operation. Its *Moving Picture Experts Group* (MPEG) header tag information and the runtime data are shown on the last line in the figure.

Project Expansion

I have shown you a basic MP3 player that can be expanded in a variety of ways to meet future requirements. The display could be changed from a standard monitor to a multiple line serial *liquid crystal display* (LCD). For now, it is sufficient to

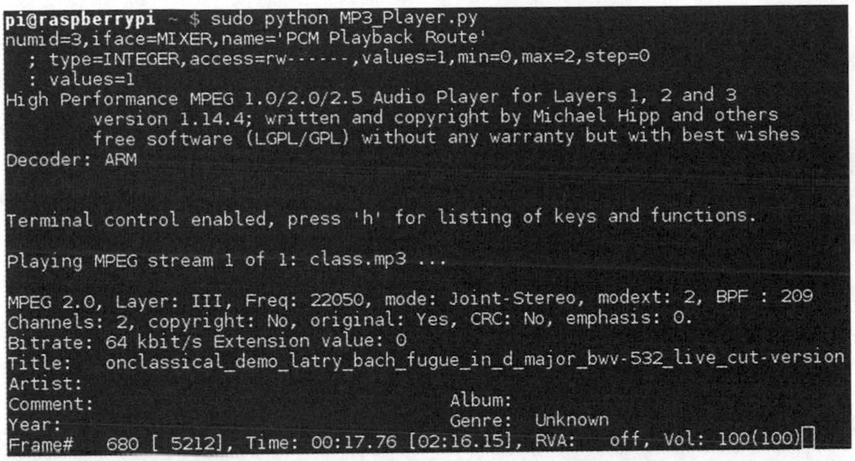

```
pi@raspberrypi ~ $ sudo python MP3_Player.py
numid=3,iface=MIXER,name='PCM Playback Route'
  ; type=INTEGER,access=rw------,values=1,min=0,max=2,step=0
  : values=1
High Performance MPEG 1.0/2.0/2.5 Audio Player for Layers 1, 2 and 3
        version 1.14.4; written and copyright by Michael Hipp and others
        free software (LGPL/GPL) without any warranty but with best wishes
Decoder: ARM

Terminal control enabled, press 'h' for listing of keys and functions.

Playing MPEG stream 1 of 1: class.mp3 ...

MPEG 2.0, Layer: III, Freq: 22050, mode: Joint-Stereo, modext: 2, BPF : 209
Channels: 2, copyright: No, original: Yes, CRC: No, emphasis: 0.
Bitrate: 64 kbit/s Extension value: 0
Title:   onclassical_demo_latry_bach_fugue_in_d_major_bwv-532_live_cut-version
Artist:
Comment:                                  Album:
Year:                                     Genre:  Unknown
Frame#   680 [ 5212], Time: 00:17.76 [02:16.15], RVA:   off, Vol: 100(100)
```

Figure 3-11 An MP3 playing.

focus on the simpler aspects of using the RasPi for project control.

Another aspect that you might have wondered about is the use of a separate push button dedicated to a specific function, such as skip to the next track. Right now, you simply press the `f` key to do this. Incorporating a unique button capability entails a lot of software rework because the mpg123 app expects control key input from what is logically known as the *stdin* or standard input device, i.e., the keyboard. Changing or paralleling the input device is not a trivial change, so I felt it was not suitable to include it in a beginning project book.

Adding MP3 songs to be played is rather easy; as all you need to do is copy them into the default directory using the RasPi's USB port. The program does have to be changed to reflect the new song names. However, the program can be modified to list all the MP3 songs in the directory so that you can select the ones to be played, which is the essence of a play list.

Of course, you can connect a battery to power the RasPi that will make it completely portable, assuming you are using an LCD-type display for user interaction.

Summary

I started this chapter with a discussion of the Pi Cobbler prototyping aide, which is an invaluable tool that allows us to experiment with most of the GPIO pins that are available on the RasPi.

The Linux ALSA system discussion showed you how sound is handled by the RasPi OS. You must install ALSA in order to use the mpg123 app that forms the basis for the MP3 player project. I also introduced you to the principal audio file formats, including the MP3 format that is used to encode the music played in the project. There was also a brief detour to discuss the RasPi's audio frequency response using a pink-noise stimulus.

I then showed you how to use push-button switches to signal the RasPi, using GPIO pins as inputs. These switches were also used as part of the MP3 player project.

We went through a thorough discussion of the mpg123 MP3 app including a detailed list of all the controls available for use with this versatile app.

I finished with a brief discussion on how to expand the MP3 player so that it might more readily resemble a modern portable player.

CHAPTER 4

Camera Controller

In this chapter, I will show you how to build a camera controller by using the RasPi with several auxiliary modules that will act as trigger sources. The controller will allow you to trigger a digital camera and/or an electronic flash if they are inherently capable of being remotely controlled. It will also enable you to take some very interesting pictures including stop-action photos.

The following three figures illustrate one simple experiment that demonstrates the controller's stop-action effect. I made a flag object out of black and white duct tape onto which I printed the numbers 1 and 2. I attached this "flag" to a nail that I inserted into a portable drill chuck. Figure 1 shows the drill chuck with the flag at rest.

Next, I started the drill and had it rotating at approximately 500 *revolutions per minute* (r/min). Figure 4-2 is a normal photograph taken with my usual studio lights and without an electronic flash. This figure mimics what the human eye perceives in that both numbers are visible because I used a slow shutter speed.

Figure 4-1 Flag at rest.

Figure 4-2 Flag rotating at 500 r/min.

Figure 4-3 is a picture taken with the electronic flash and a slow shutter speed in a darkened room.

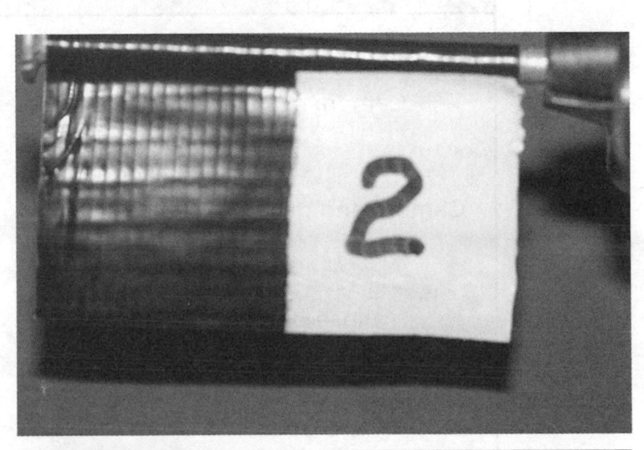

Figure 4-3 Flag captured in a stop-action photo.

You can clearly see the number 2 visible and not blurred, even though it was rotating at 500 r/min or almost 10 times per second.

The controller will assist you in making similar stop-action pictures as well as many other effects that you may want to accomplish.

Project Requirements

I will start this project by listing the requirements that the controller system should meet. These have been simplified to make the controller compatible with a wide variety of cameras and electronic flash units. In order to take full advantage of all the system functions, you will need a camera that is capable of being remotely triggered. The controller system requirements are shown in Table 4-1.

Figure 4-4 is a high-level graphical diagram showing the module interconnections associated with each mode listed in Table 4-1.

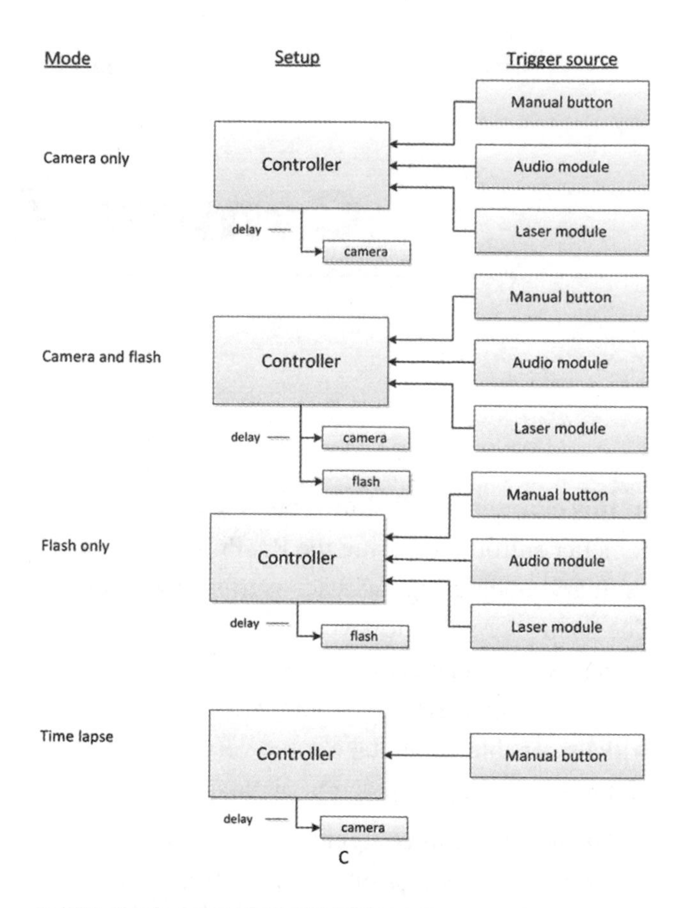

Figure 4-4 High-level camera controller diagram.

Camera Shutter Operation

A brief discussion on how a *Digital Single Lens Reflex* (DSLR) shutter operates will help you understand the interface between the RasPi and the camera. Modern DSLRs, such as the Canon 40D that I am using in this project, have a multifunction shutter control. The camera will autofocus when the shutter button is depressed halfway and operate the shutter with a full press. Figure 4-5 shows the

Table 4-1 Camera Controller Requirements

Mode	Trigger Source	Delay Range	Shutter Operation
Camera only	Manual Sound module Laser module	0–10 seconds 1 msec resolution	Focus (half press) Shutter (full press)
Camera and flash	Manual Sound module Laser module	0–10 seconds 1 msec resolution	Focus (half press) Shutter (full press)
Flash only	Manual Sound module Laser module	0–10 seconds 1 msec resolution	N/A
Time lapse	Manual	N/A	0–8 hours 1 second resolution Shutter (full press)

Figure 4-5 Canon 40D shutter-control connector.

interface connector that has both the focus and shutter-control pins along with the common or ground pin. There is 3.3 V present on both the focus and control pins, and connecting either pin to ground completes the circuit and causes its corresponding action (focus or control) to be actuated.

A special plug has to be used with this connector. I cannibalized a commercial, inexpensive, remote shutter control in order to have a plug for this project. Figure 4-6 shows this plug with a molded cable.

You can go to the website www.doc-diy.net/photo/remote_pinout/ to look up your camera's shutter-control pin layout and matching connector. Some of the newest DSLRs are using a standard

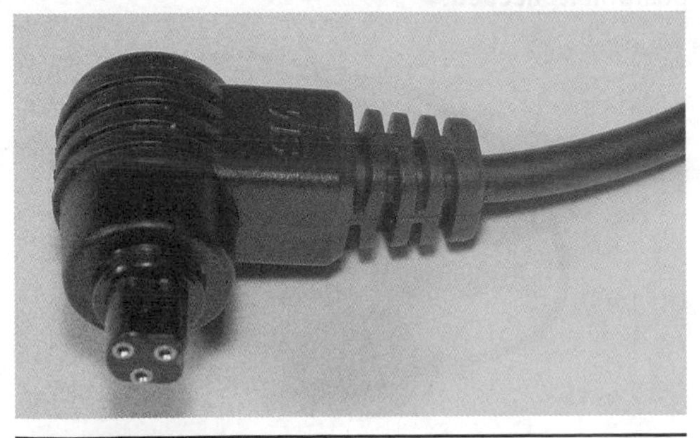

Figure 4-6 Canon shutter-control plug and cable.

3.5-mm stereo jack that will make the interface connection much easier.

Do not despair if you do not have a camera capable of being remotely controlled. Simply use the sound or light-activated functions to trigger an electronic flash with your point-and-shoot camera set to a long exposure in a darkened room. The flash will illuminate the subject sufficiently to enable you to do stop-action photography. You will have to experiment with different exposure times to get the desired result, but you do have the basic equipment to be successful.

DSLR owners will learn how to activate the focus and shutter controls in the optoisolator section after reading the electronic flash discussions, which are next.

Electronic Flash

The electronic flash, or strobe, is the unit that produces a very bright light for a very brief instant in time. It is used to sufficiently illuminate a photographic subject where ambient light is too low in intensity or not suitable for the subject. The electronic flash replaced the one-time-use flash bulb many years ago and is now almost universally used in modern photography. Most digital cameras, from one-steps to DSLRs, have a built-in flash unit that automatically activates based upon the lighting conditions that the camera sensor detects. While this arrangement may suit most users, for this project, we cannot make use of the built-in unit but instead must rely on an external electronic flash that can be remotely triggered. Figure 4-7 shows the Canon 420EX electronic flash that I will use in this project. It is several years old and has been replaced by the model 430EX. For our purposes, however, the differences between the models are not significant.

The 420EX is designed to mount into the camera's hot-shoe connection that provides the shutter activation as well as a series of data interconnections. The data flow between the flash

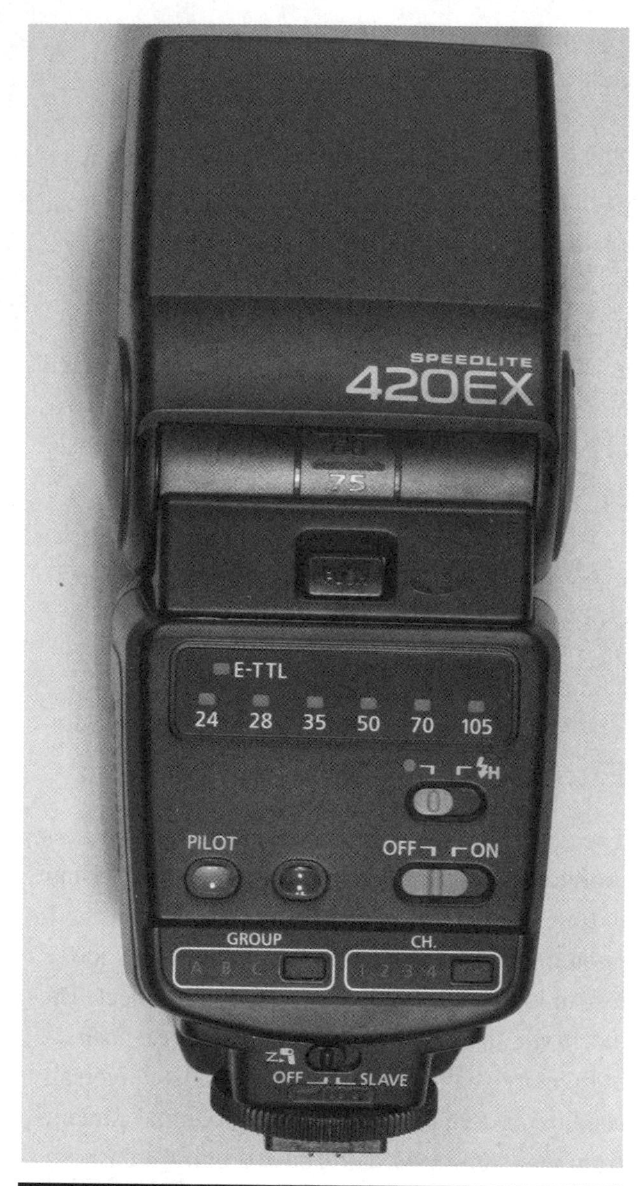

Figure 4-7 Canon model 420EX electronic flash.

and the camera is not relevant to the project, as the flash will not be mounted on the hot shoe. However, it is important to point out the control connections that will be used to trigger the flash. These are detailed in Fig. 4-8.

The flash will fire when the shutter-control pin is connected to the ground pin.

NOTE I used the descriptor, ***shutter control,*** to emphasize that the flash is fired only when the camera shutter operates. This pin obviously triggers the flash itself.

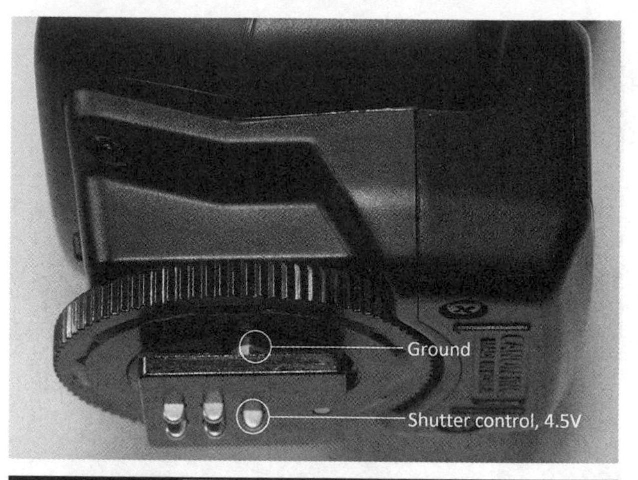

Figure 4-8 Flash hot-shoe connections.

The ground pin, in actuality, consists of two spring-loaded, cam-shaped pins located on either side of the inner slides for the hot shoe. These make contact with the camera's metal hot shoe. There are five spring-loaded pins shown on the bottom of the hot shoe. As mentioned earlier, we are only concerned with the center pin. The other four pins visible to the left of the center pin are data connections that are not used in this project. I also used an inexpensive, commercially available hot-shoe adapter to bring out both the shutter-control and the ground connections. This adapter is shown in Fig. 4-9.

The adapter brings out the connections to a 3.5-mm plug that can be inserted into a standard mono jack. The adapter also incorporates a standard tripod mount, as shown in the figure inset. This makes it simple to mount the flash in any position as needed.

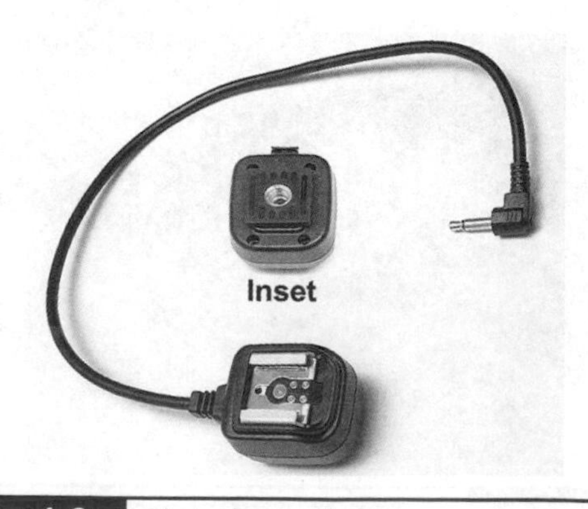

Figure 4-9 Hot-shoe adapter.

Electronic Flash Control Voltages

You may have noticed that I put the 4.5-V control voltage in the Fig. 4-8 pin description. This is a typical voltage found on Canon electronic flash units, and it is quite compatible with this project's interface circuits. However, older electronic flash units may have significantly higher voltages present on the control pin. These may range up to several hundred volts.

CAUTION **Connecting an electronic flash with more than 24 V on the trigger contact to the project interface circuit will destroy the interface and may cause further damage to the RasPi.**

Modern electronic flash units are expected to conform to ISO Standard 10330:2002 entitled "Photography—Synchronizers, ignition circuits and connectors for cameras and photoflash units—Electrical characteristics and test methods." One portion of the standard requires that trigger voltages not exceed 24 V max. Canon has designed their units to operate at less than 6 V, and Nikon at less than 12 V. Older units were not designed to the standard and could conceivably have hundreds of volts exposed that would not only destroy the interface but also present a personal shock hazard. To check flash voltages for your electronic flash, go to the following excellent website: www.botzilla.com/photo/strobeVolts.html.

I do discuss a possible solution for using an older flash unit in the next section on the optoisolator interfaces. However, my earnest suggestion would be: Do not use an old flash, but instead invest in a modern unit. Many of these are available at quite reasonable prices.

Optoisolator Interfaces

An optoisolator combines a light source with a photosensitive transistor to create a photonic coupler. This type of isolator achieves electrical

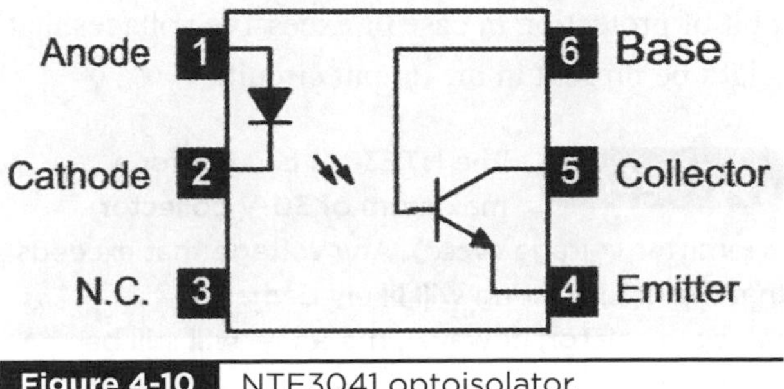

Figure 4-10 NTE3041 optoisolator.

isolation between two electrical circuits that may have a significant voltage difference present. Figure 4-10 shows a physical schematic for the NTE3041 optoisolater that is used in this project. It is in a six-pin *dual in-line package* (DIP) form with a gallium-arsenide infrared LED as a light source and a silicon NPN phototransistor as the light receiver.

Although it might appear that an optoisolator could pass analog or continuous signals, it is only suitable for digital or on/off signals. That is due to the on/off functional nature of the LED used as the light source. NTE3041s are used as input and output isolators in this project. Figure 4-11 shows a very simple circuit that is used to enable the NTE3041 to isolate one of the RasPi GPIO pins. All that is needed is a current-limiting resistor in series with the GPIO pin. The resistor value was set at 220Ω to provide a 10-mA drive current to the LED in the optoisolator. This current level is suitable for both the RasPi and the optoisolator. There is also a 3-V Zener diode connected in series with the photo transistor's collector, which affords

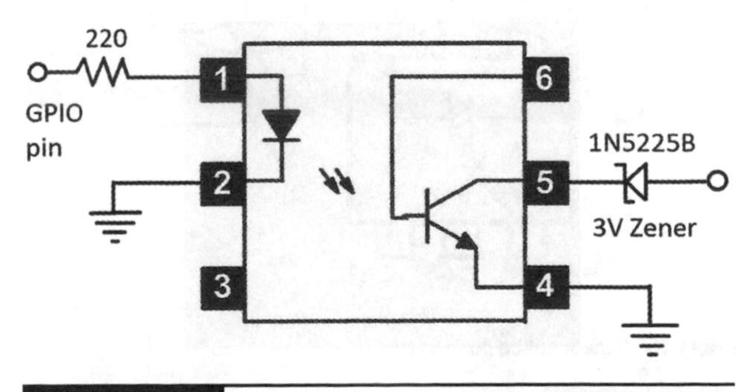

Figure 4-11 RasPi to optoisolator circuit.

a bit of protection in case of excessive voltages that might be present in the output circuit.

CAUTION **The NTE3041 is rated for a maximum of 30-V collector-to-emitter voltage (V_{CEO}). Any voltage that exceeds that maximum rating will likely damage the optoisolator; however, the RasPi will still be protected.**

I have used the NTE3041 optoisolators to connect the RasPi GPIO pins to the shutter (full press), focus (half press), and electronic-flash control pins. The sound and light modules do not require isolation, as they already use isolated relay contacts.

High-Voltage Optoisolator Circuit

Figure 4-12 shows a circuit that can be used if you need to control an electronic flash that has high voltage present on the control pins. This circuit uses an optotriac (*triode for alternating current*) isolator in lieu of the NTE3041 optoisolator. The optotriac, in turn, turns on a *silicon-controlled rectifier* (SCR) that fires the flash unit. The SCR recommended is rated up to 400 V. I strongly recommend you do not use a flash unit with more than that voltage.

Connect the RasPi GPIO pins to pins 1 and 2 of the optotriac in the same manner as shown in the NTE3041 connections of Figure 4-11. Be sure to include the 220 Ω current limiting resistor.

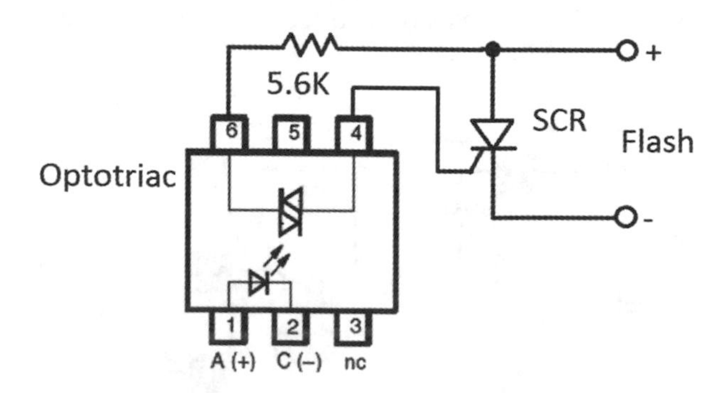

Figure 4-12 High-voltage optoisolator circuit.

Table 4-2 High-Voltage Optotriac Parts List

Part	Model Number	Source
Optotriac	Motorola MOC3010M	Mouser
SCR	STM 511-TYN408	Mouser
5.6 kΩ	½ watt	

Information about the suggested parts for the high-voltage optoisolator circuit is provided in Table 4.2.

Sound and Laser Modules

An important part of this project is to have the capability of triggering the camera and flash by using sound and light events. This means we need sensors to detect the sound and light as well as the associated circuitry to generate a signal when the preset thresholds are exceeded. These requirements raise an issue that often comes up in project planning: whether to build or buy. In this case, I decided to incorporate two inexpensive kits for sound and light detection. My decision was based on an estimate of how long it would take to build and test two prototypes versus buying and building two kits and making any appropriate modifications to suit this project. Listed here are the kits I used:

Global Specialties Model GSK-409 Audio Control Switch kit $9.95

Ramsey Electronics Model LTS1 Laser Trip Sensor kit $19.95

It turned out that neither kit required any interface modifications for use in the project, since both had output relays that could directly connect with the RasPi GPIO pins without the need for optoisolators.

Audio Module

An assembled audio module is shown in Fig. 4-13. You will notice that I connected the microphone to the main board with a foot-long twisted-wire

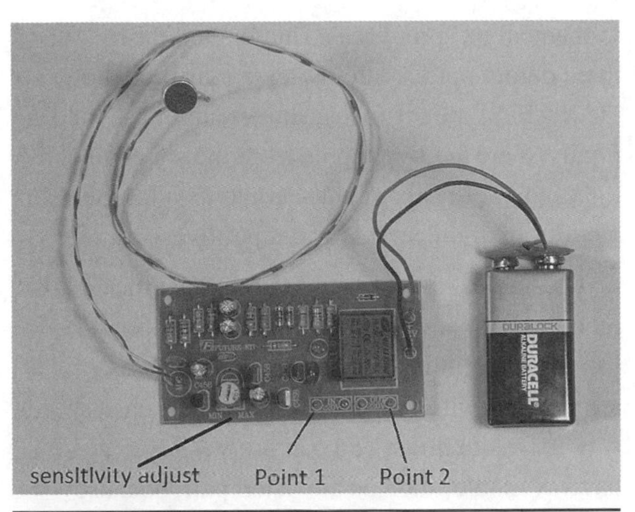

Figure 4-13 Model GSK-409 audio control switch.

pair, so that I would be able to position it for optimal sound pickup. The microphone is normally soldered directly to the main board, which would have made it awkward to position the board with the 9-V battery attached in order to best detect the trigger sound.

This board does feature a flip-flop operation where the output relay is latched after a sound event. This means that relay contacts will close and remain closed after the first sound event. The next sound event will open the relay contacts, and they will stay that way until the next event, ad infinitum. This is really not a limiting feature for the project, since you would probably have planned to manually reset the module before any event using sound as a trigger.

Notice the sensitivity trim pot that I have pointed out in Fig. 4-13. I found that I had to adjust the pot almost fully clockwise, in order to have a reasonable opportunity to detect low-to-moderate volume sound events.

This module is easily connected to the Pi Cobbler prototype board using the schematic shown in Fig. 4-14.

Point 1 shown in Figs. 4-13 and 4-14 is connected to the Pi Cobbler ground pin. Point 2 also shown in Figs. 4-13 and 4-14 is connected to GPIO pin 24 on the Pi Cobbler. There is also a

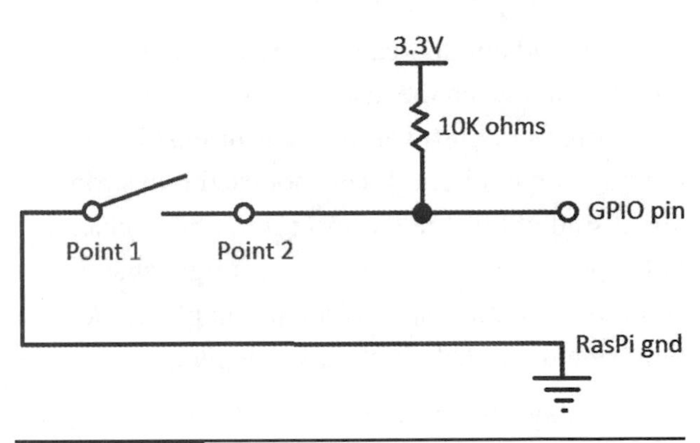

Figure 4-14 Sound module to Pi Cobbler interface schematic.

10-kΩ pull up resistor tied to the GPIO pin. The RasPi control program will be written to detect the high-to-low transition on the selected GPIO pin.

Laser Module

An assembled laser-light module is shown in Fig. 4-15. There are two assemblies, not counting the 12-V power supply, that make up this sensor module. The first is the laser pointer that acts as the light source, and the second is the light sensor that consists of a phototransistor with some analog signal processing circuitry. The laser pointer is a very common low-power device that is typically battery powered. In this case, it is powered by 5 V supplied by the detector board. The power is delivered through a pair of small alligator clips, one of which is visible in the Fig. 4-15.

The light detector component is a phototransistor that is also pointed out in the figure. It looks like a normal LED but is a transistor with only the

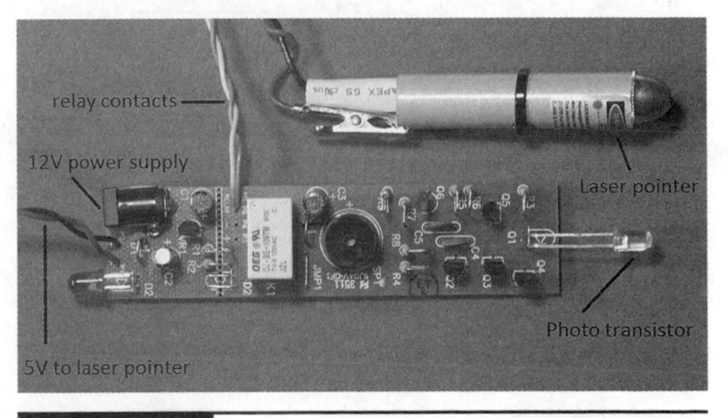

Figure 4-15 Model LTS1 laser trip sensor.

collector and emitter leads externally connected. It will conduct when the laser light strikes the base-emitter region; however, it seems insensitive to normal ambient light. I have not tried it in strong direct sunlight where I suspect it might conduct. I also purposely left off an ambient light shield to show you the transistor. When put in place, the light shield will help with false triggers.

The relay contacts in this module are normally closed when light is not being detected and will open when the laser light strikes the phototransistor. This means that the RasPi program will need to be changed to detect a low-to-high transition provided the same interface is used, as is shown in Fig. 4-14. That's what is so nice about using the RasPi and Python: changes can easily be made in software to accommodate hardware issues.

Interface Connections

All of the interface connections are shown in Figure 4-16. There are three input push buttons to manually trigger focus, shutter, and flash; three output optoisolator connections that allow the RasPi to trigger focus, shutter, and flash; and finally, there are two inputs from the sound and light modules. This might seem to be a lot, but the circuits are quite simple and repetitive.

The hardest part was to interconnect the modules, camera, and flash to the breadboards. I finally determined that using audio cables terminating with 3.5-mm jacks was the easiest way for me to do it. You can use whatever you have available, such as a twisted pair, because the connections are DC based with practically no speed or bandwidth issues. The complete prototype is shown in Fig. 4-17. It is a bit of a "rat's nest" as far as the wiring goes, but it is totally functional and allowed me to rapidly progress with the project.

CAUTION **Ensure that you use the appropriate ground connections, as shown in Fig. 4-16. For instance, the camera ground is separate and distinct from the RasPi**

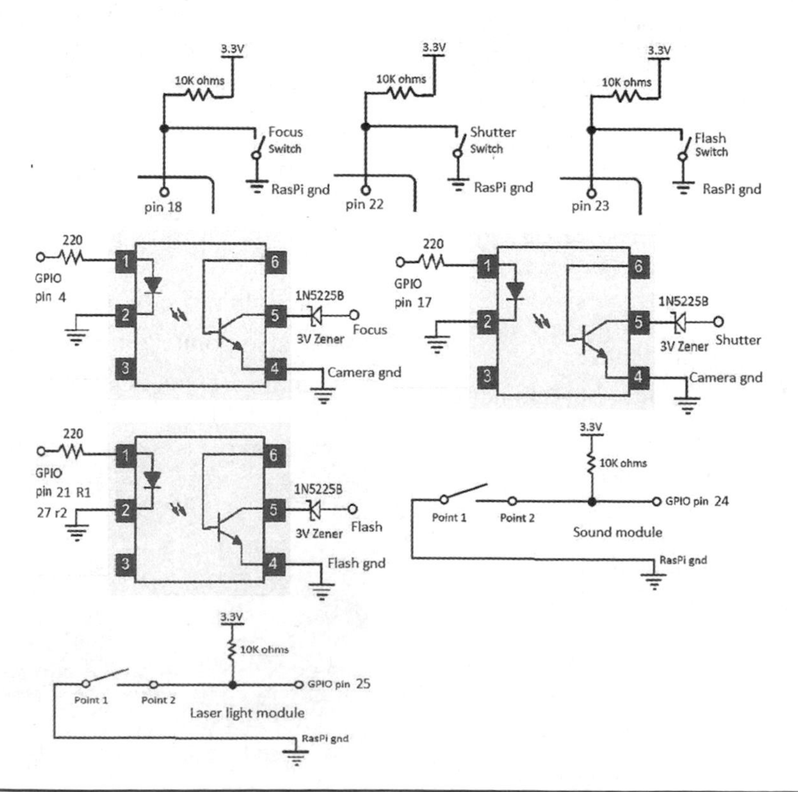

Figure 4-16 All of the interface connections.

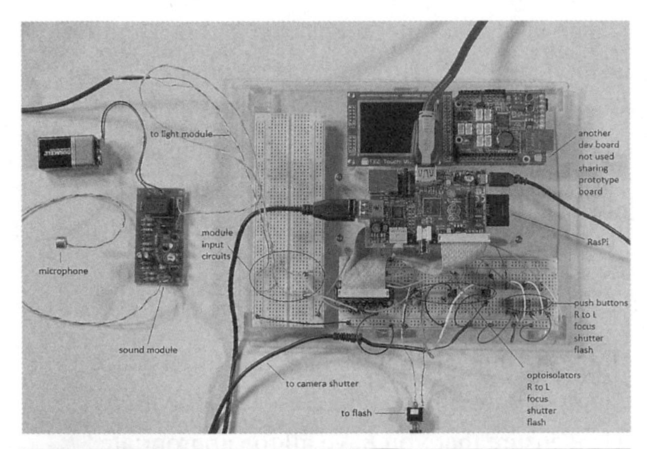

Figure 4-17 Complete camera controller prototype.

ground. No harm would result if you inadvertently connected the optoisolator grounds to the breadboard ground, but the isolators would simply not function properly.

Table 4-3 lists all the GPIO pins used in the interface with the associated functions tied to those pins.

NOTE **Starting with revision 2 for Models A and B, pin 21 has been redesignated pin 27. It remains pin 21 for revision 1 models A and B. The Pi Cobbler, Pi Plate, and likely other prototype tools have pin 21 shown. I would suggest putting a piece of tape over the 21 and relabeling it 27 if you are using revision 2 models. The GPIO library recognizes pin 27 without any issues.**

Testing the Interface Connections

You should test all the connections to ensure that everything is wired correctly. The only item you will need is a standard *volt-ohm meter* (VOM) to run through all the tests. You will not require any software to be running on the RasPi for these tests although the RasPi must be turned on in order to provide power to the Pi Cobbler and breadboards.

First check each of the push buttons. There should be 3.3 V present on pins 18, 22, and 23. The voltage should drop to 0 V when one of the buttons connected to the pin under test is pushed.

Next, check the camera optoisolators. First ensure that the wires going from pins 4, 17, and 21 are temporarily disconnected. Then plug the shutter-control cable into the camera and then turn on the camera. Using a jumper wire, connect the 3.3-V power supply on the breadboard to the 220-Ω resistor that connects to pin 1 of the optoisolator.

CAUTION **Do not touch pin 1 on the optoisolator with the 3.3-V jumper wire.** **You must go through the current-limiting resistor or else the LED in the optoisolator could burn out from excessive current.**

You should observe either the camera auto-focusing or the shutter operating, depending upon which optoisolator you are testing.

Table 4-3 GPIO Pin Function Interface List

Pin number	Input/Output	Function
4	Output	Focus optoisolator
17	Output	Shutter optoisolator
18	Input	Focus push button
21 (revision 1)	Output	Flash optoisolator
22	Input	Shutter push button
23	Input	Flash push button
24	Input	Sound module input
25	Input	Laser-light module input
27 (revision 2)	Output	Flash optoisolator

The electronic flash should be tested next. Plug the remote hot shoe into the flash and ensure that it is also connected to its controlling optoisolator. Turn on the flash and then touch the 3.3-V jumper wire to the 220-Ω, current-limiting resistor that is connected to the optoisolator. The flash should operate.

Remember to reconnect all the wires between the optoisolators and their respective GPIO pins, as shown in Fig. 4-16.

The sound and light modules are tested last. Connect the modules to the breadboard and power them on. You do not have to disconnect the wires going between the GPIO pins and the module interfaces for these tests. The sound module connection will be either on or off, depending upon the state of the latched relay. Using the VOM, check to see that pin 24 switches between 3.3 V (on) and 0 V (off). This is a valid test for the sound module because it operates in the same manner as the push-button interface.

The laser-light module is checked in a similar fashion at pin 25. However, as mentioned above, the module-relay contacts are closed when light is not striking the phototransistor. This means that pin 25 should have 0 V when no light is on the sensor and 3.3 V when light strikes the sensor.

If you have any problems, recheck all the wiring, since it is very easy to misplace a jumper wire on the solderless breadboards. Also, ensure that you have all the appropriate grounds connected, as shown in the figures.

Initial Test Program

The following program tests the camera focus and shutter controls as well as the electronic flash by using the push buttons. It is named CameraControl .py and may be downloaded from the book's website, www.mhprofessional.com/raspi.

CameraControl.py

```
import time
import RPi.GPIO as GPIO
GPIO.setmode(GPIO.BCM)
GPIO.setup(4, GPIO.OUT)
GPIO.setup(17, GPIO.OUT)
#I am using a rev 2 board, so I will use pin 27 in lieu of pin 21.
GPIO.setup(27, GPIO.OUT)
GPIO.setup(18, GPIO.IN)
GPIO.setup(22, GPIO.IN)
GPIO.setup(23, GPIO.IN)
GPIO.setup(24, GPIO.IN)
GPIO.setup(25, GPIO.IN)
while True:
    #check for the focus push-button press.
    if(GPIO.input(18) == False):
        #trigger the optoisolator connected to the focus lead.
        GPIO.output(4, GPIO.HIGH)
        #need a little time for the camera to focus.
        time.sleep(.2)
        #reset the optoisolator.
        GPIO.output(4, GPIO.LOW)
```

```
    #check for the shutter push-button press.
    if(GPIO.input(22) == False):
        #trigger the optoisolator connected to the shutter
        lead.
        GPIO.output(17, GPIO.HIGH)
        time.sleep(.2)
        GPIO.output(17, GPIO.LOW)
    #check for the flash push-button press.
    if(GPIO.input(23) == False):
        #trigger the optoisolator connected to the flash. (I use pin 27 for rev 2
        board.)
        GPIO.output(27, GPIO.HIGH)
        time.sleep(.2)
        GPIO.output(27, GPIO.LOW)
```

I recommend running this program directly from a terminal window as follows:

```
sudo python CameraControl.py ↵
```

Press each of the push buttons and confirm that the appropriate action occurs. When you are finished, type `^c` to exit the program.

You should note that I reset the controlling GPIO pin after making it a HIGH value. This ensures that it is in the proper state for the next operation.

Sound Module Test Program

The following program tests the sound module to check that the focus, shutter, and flash can be triggered when the sound module detects a sound event. The first portion of this program is identical to the previous program. This program interacts with you through the screen by prompting a number representing the test to be performed. The prompts are shown on the screen and are also shown here:

1—Test the shutter only.

2—Test the flash only.

3—Test the shutter and flash together.

0—Exit the program.

The program is coded as a series of loops that wait for a trigger event from the sound module before activating the requested function(s). This program is available on the book's website as SoundModTest.py.

Just be aware that you must reset the sound module after it has been triggered because of the latched operation. It is a simple matter of a one-hand clap to trigger it so that the module LED turns off and the module is in a proper ready state.

Light Module Test Program

The program to test the light module is almost identical to the sound-module test program.. The only changes are in the three `GPIO.input` conditional `if` statements that need to be changed from `if(GPIO.input(24) == False) to if(GPIO.input(25) == True)`. These changes are required because the light module is connected to pin 25 in lieu of pin 24 and its relay contacts are normally closed instead of normally open, as I mentioned earlier in this chapter.

To save book space, I will not repeat the code listing, as the changes are minor. This program, SoundModTest.py, is available from the book's website and is named LightModTest.py.

SoundModTest.py

```
import time
import RPi.GPIO as GPIO
GPIO.setmode(GPIO.BCM)
GPIO.setup(4, GPIO.OUT)
GPIO.setup(17, GPIO.OUT)
#I am using a rev 2 board, so I will use pin 27 in lieu of pin 21.
GPIO.setup(27, GPIO.OUT)
GPIO.setup(18, GPIO.IN)
GPIO.setup(22, GPIO.IN)
GPIO.setup(23, GPIO.IN)
GPIO.setup(24, GPIO.IN)
GPIO.setup(25, GPIO.IN)
select = input('Enter 1 (shutter only), 2 (flash only), 3 (shutter and flash), 0
(exit): '):
while select > 0:
    # shutter only.
    if select == 1:
        while True:
            if(GPIO.input(24) == False):
                GPIO.output(17, GPIO.HIGH)
                time.sleep(.2)
                GPIO.output(17, GPIO.LOW)
                break
    # flash only.
    if select == 2:
        while True:
            if(GPIO.input(24) == False):
                GPIO.output(27, GPIO.HIGH)
                time.sleep(.2)
                GPIO.output(27, GPIO.LOW)
                break
    # shutter and flash.
    if select == 3:
        while True:
            if(GPIO.input(24) == False):
                GPIO.output(17, GPIO.HIGH)
                GPIO.output(27, GPIO.HIGH)
                time.sleep(.2)
                GPIO.output(17, GPIO.LOW)
                GPIO.output(27, GPIO.LOW)
                break
    select = input('Enter 1 (shutter only), 2 (flash only), 3 (shutter and flash), 0
    (exit): '):
print('All done')
```

To test this module you will need to aim the laser at the phototransistor, which should turn off the LED on the detector board. Next, block the laser light beam with an opaque item, which should trigger the light module and initiate the requested function. There is no need to reset the light module, since it is not latched.

Time-Lapse Function

The last camera controller function to test is time-lapse photography. This is where you set a maximum time to take photos at a preset interval. I created the program using only the shutter function, as I believe that will be its primary use. Normally, time-lapse photography is done outdoors without the need of a flash. Also, it doesn't make sense to incorporate the sound or light triggers for this function. The program is quite simplified as compared to the previous programs and is listed below. It is also available from the book's website as TimeLapse.py.

NOTE The interval should be entered as a decimal if you need fractions of a minute, i.e., an interval of six seconds would be entered as .1.

Operational Code

The four test programs, CameraControl.py, SoundModTest.py, LightModTest.py and TimeLapse.py should serve as templates to create whatever style operational program meets your needs. It is impossible for me to create the wide variety of control programs that are possible because there are so many different environments that would need to be accommodated. It is easy to add additional functionality to any of these programs by simply adding the desired input and/or output triggers. For instance, if you needed a flash for the time-lapse function, you would add `GPIO.output(27, GPIO.HIGH)` after the `GPIO.output(17, GPIO.HIGH)` statement and do a similar add after the GPIO reset. Also note that I used 27 in lieu of 21 due to the rev 2 board I used.

TimeLapse.py

```
import time
import RPi.GPIO as GPIO
GPIO.setmode(GPIO.BCM)
GPIO.setup(17, GPIO.OUT)
max = input('Enter maximum time (minutes): ')
interval = input('Enter interval time (minutes): ')
# need a variable to keep track of total elapsed time.
elapsed = 0
# keep looping until elapsed time exceeds max time.
while elapsed < max
    # the sleep function argument is in seconds so we must multiply by 60.
    time.sleep(60 * interval)
    # shutter only operation.
    GPIO.output(17, GPIO.HIGH)
    time.sleep(.2)
    GPIO.output(17, GPIO.LOW)
    elapsed = elapsed + interval
print('All done')
```

Summary

This chapter started with a project-requirements list that also served as a list of goals that you needed to accomplish. Requirements drive all projects, and you should spend a great deal of time analyzing and determining what the essential requirements or needs are before proceeding to any design work.

I next discussed how a DSLR camera with an electronic flash could be remotely controlled. I elected to use optoisolators where appropriate to lessen any chance of damaging voltages being introduced into the RasPi interface circuits. I also mentioned that you could use a non-DSLR camera, set to a long exposure, with the electronic-flash function to capture stop-action events.

In a detailed discussion regarding optoisolators, I explained that they are very handy for completing interface circuits that incorporate a high level of voltage isolation. I also provided a triggering circuit for those brave souls with a high voltage electronic flash.

I went through my build-versus-buy decision regarding the sound and light modules. I was very pleased to find that I had made the right decision, as I had no trouble incorporating these kits into the project. You will find that your time is very valuable and finding ways to shorten a project, such as incorporating prebuilt components and/or kits, can be a valuable tool for maximizing your productivity.

I next went through a step-by-step testing procedure to ensure that the interface circuits were working as designed. There was no need to run any programs on the RasPi at this stage. It is always a prudent idea to carefully check and test all interface circuits before running any programs. Testing in this fashion allows you to determine that any later functional problems are probably due to program glitches, not the interface circuits.

I then introduced a series of four test programs to methodically check that all planned project requirements were met and that all the components functioned as desired. These four programs will serve as good templates to allow you to create your own custom programs. You can design them to best suit your needs, or you can simply use them as they are.

CHAPTER 5

GPS

Introduction

In this chapter, I will first discuss what constitutes the *Global Positioning System* (GPS), a satellite-based navigation system, and the advantages and limitations that are involved with using it. I will next explain how a serial communication link may be set up and run between the RasPi and a laptop computer.

Finally, I will show you how to connect a very capable GPS receiver to the RasPi and subsequently use a series of applications to decode and display *target, position, and velocity* (TPV) information on the RasPi screen.

Brief GPS History

The GPS system was initially deployed in the early 1970s by the U.S. Department of Defense (DoD) to provide military users with precise location and time synchronization services. Civilian users could also access the system; however, the services offered to the public were purposely degraded to avoid national security risks. The DoD was concerned that the system might help potential enemies in their activities. This intentional degradation was lifted by order of President Reagan in 1983 to allow civilian use of full and more accurate GPS services. In 2000, an even higher level of accuracy without any degradation was implemented for all users.

The current GPS system has 32 satellites in high orbits over the earth. A representative diagram of the satellite "constellation" is shown in Fig. 5-1. The satellite orbits have been carefully designed to allow for a minimum of six satellites to be in the instantaneous field of view of a GPS user located anywhere on the surface of the earth. A minimum of four satellites must be viewed in order to obtain a location fix, as you will learn in "The Basics of How a GPS Functions" section.

Several other GPS systems are also deployed:

GLONASS—The Russian GPS

Galileo—The European GPS

Compass—The Chinese GPS

IRNSS—The Indian Regional Navigation Satellite System"

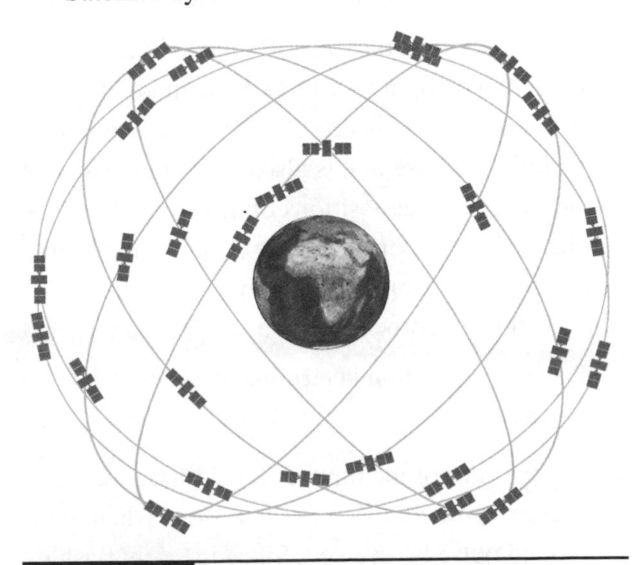

Figure 5-1 Diagram of the GPS satellite "constellation."

I will be using the American GPS system, as vendors have made many inexpensive receivers available to purchase. All receivers function in essentially the same way and conform to the *National Marine Electronics Association* (NMEA) standard discussed in the "NMEA Protocol" section.

The Basics of How GPS Functions

I made up an analogous fictional position-location system to help explain how the GPS system functions. First, imagine a two-mile by two-mile land area where this system is set up. The land terrain contains gently rolling hills, each no more than 30 feet in height. The subject, using a "special" GPS receiver, may be located anywhere within this area. Also located in this area are six 100-foot towers, each containing a beacon. The beacon atop each tower is configured to briefly flash a light and emit a loud sound burst simultaneously. Each beacon also emits light and sound pulses once a minute but at a specific time within the minute. Beacon one (B1) emits at the start of the minute, beacon two (B2) at 10 seconds past the start of the minute, beacon three (B3) at 10 seconds later, and so on for the remaining beacons.

It is also critical that the GPS receiver have an open line of sight to each beacon and that the position of each beacon be recorded in an embedded database that is also constantly available to the receiver. The positions of beacons B1 through B3 are recorded as x and y coordinates in terms of miles from the origin that is shown in the upper left hand corner of Fig. 5-2.

The actual position determination happens in the following manner:

- At the start of the minute, B1 flashes, and the receiver starts a timer that stops when the sound pulse is received. Since the light flash is essentially instantaneous, the time interval is

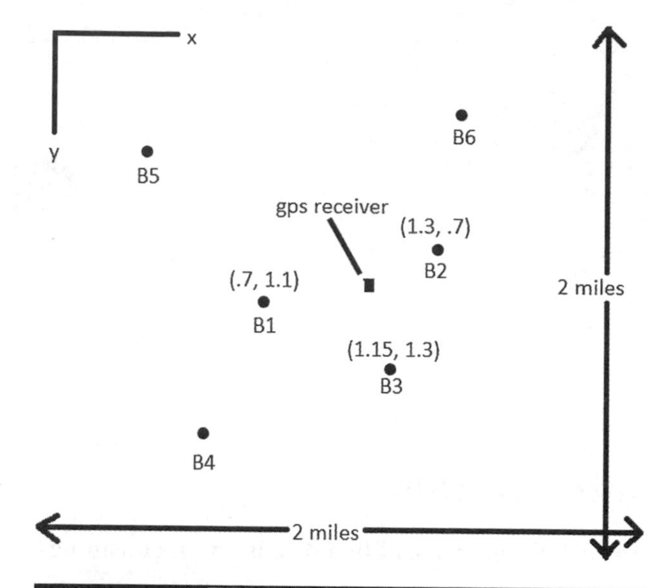

Figure 5-2 Beacon test area.

proportional to the distance from the beacon. Since sound travels at a nominal 1100 feet per second (ft/s) in air, a 5-second delay would represent a 5500-foot distance. The receiver must then be located somewhere on a 5500-foot radius sphere that is centered on beacon B1. Figure 5-3 illustrates this abstraction as a graphical representation taken from a Mathworks Matlab application.

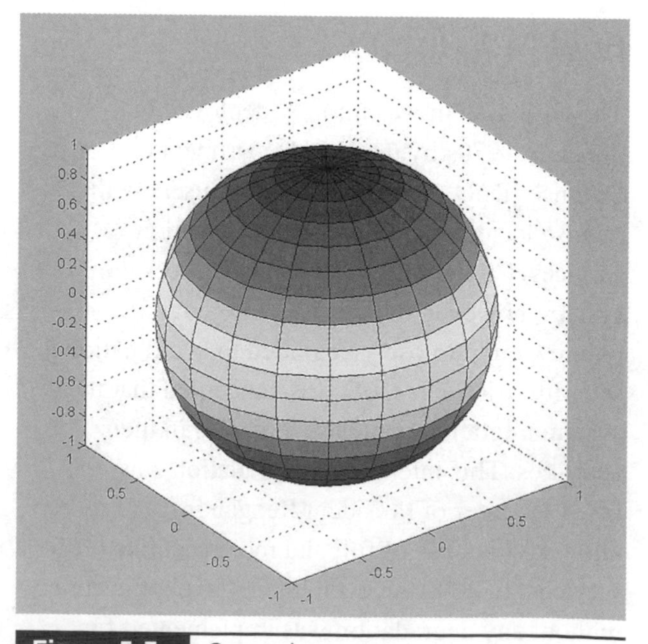

Figure 5-3 One sphere.

- B2 flashes next. Suppose that it takes 4 seconds for the B2 sound pulse to reach the GPS receiver. This delay represents a 4400-foot sphere centered on B2. The B1 and B2 spheres are shown intersecting in Fig. 5-4. The heavily dashed line represents a portion of the circle that is the intersection of these two spheres. The receiver must lie somewhere on this circle,which appears to be a straight line when observed in a planar or perpendicular view. However, there is some uncertainty about where the receiver is located on the circle. Thus, another beacon is still needed to resolve the uncertainty.
- B3 flashes next. Suppose that it takes 3 seconds for the B3 sound pulse to reach the GPS receiver. This delay represents a 3300-foot sphere centered on B3. The B1, B2, and B3 spheres are shown intersecting in Fig. 5-5. The receiver must be located at the star shown in the figure. In reality, it could be at either a high or low point, since the third sphere intersects the two other spheres at two points. The receiver position has now been fixed with regard to x and y coordinates but not the third or z coordinate. Guess what? You now need a fourth beacon to resolve whether the receiver is at the high or low point. I am not going to go through the whole process again because I think you have figured it out by now.
- Figure 5-5 shows a plane view of all three spheres with the GPS receiver position shown. You can think of it as a horizontal slice taken at $z = 0$ in Fig. 5-6.

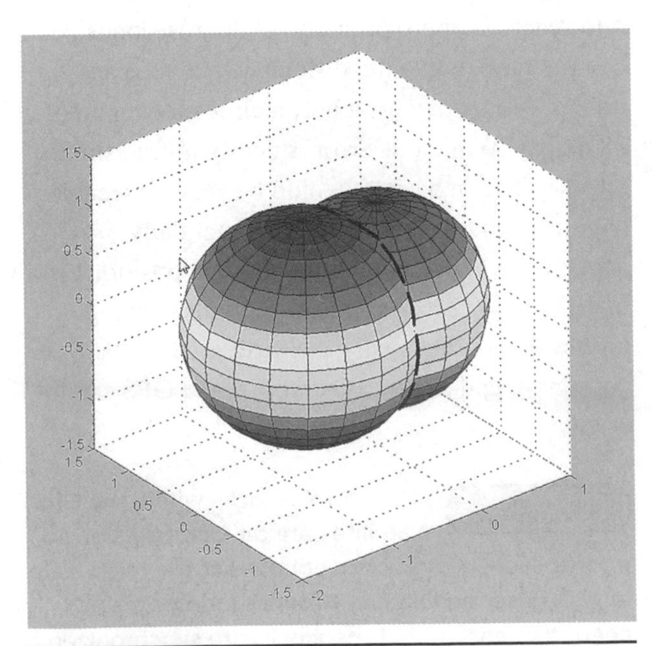

Figure 5-4 Two spheres.

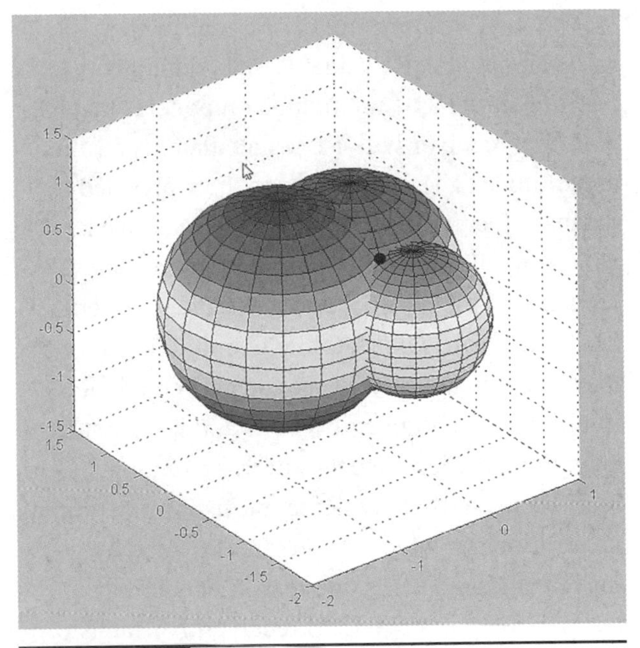

Figure 5-5 Three spheres.

In summary, it takes a minimum of three beacons to determine the x and y coordinates, and a fourth beacon to fix the z coordinate. Now translate the beacons to satellites and the x, y, and z coordinates to latitude, longitude, and altitude, and you have the basics of the real GPS system.

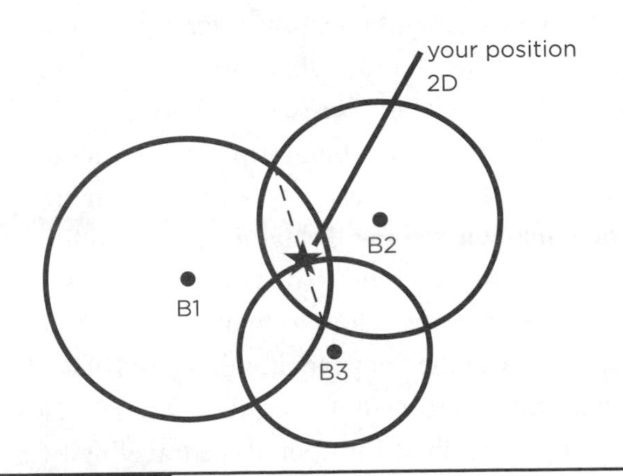

Figure 5-6 Plane view.

The satellites transmit digital microwave *radio-frequency* (RF) signals that contain both *identification* (ID) and *timing* components that a real GPS receiver will use to calculate its position and altitude. The counterpart to the embedded database mentioned in my fictional example is called an *ephemeris* or *celestial almanac*, and it contains all the data necessary for the receiver to calculate a particular satellite's orbital position. As mentioned in the history section earlier, all GPS satellites are in high earth orbits and are constantly changing position. Because of this situation, the receiver must use a dynamic means for determining its position fix, which, in turn, is provided by the ephemeris. This is one reason why it may take a while for a real GPS receiver to establish a lock, as it must go through a large amount of data calculations to determine the actual satellite positions within its field of view.

In my fictional example, the radii of the "location spheres" are determined by the receiver using extremely precise timing signals that are contained in the satellite transmissions. Each satellite contains an atomic clock to generate these clock signals. All satellite clocks are constantly synchronized and updated from earth-based ground stations. These constant updates are needed to maintain GPS accuracy, which would naturally degrade because of two relativistic effects. The best way to describe the first effect is to retell the paradox of the space-travelling twin.

Imagine a set of twins (male, female—it doesn't matter), one of whom is slated to take a trip on a fast starship to our closest neighboring star, Alpha Centauri. This round trip will take about ten years traveling nearly at the speed of light. The remaining twin will stay on earth awaiting the return of his/her sibling. The twin in the starship will accelerate very close to light speed and will patiently wait the ten years it will take to make the round trip according to the clock in the ship. Now, according to Albert Einstein, if the traveling twin could view a clock on earth he/she would observe time going by at a much quicker rate then it was in the starship. This effect is part of the *theory of special relativity* and, more specifically, is called *time dilation*. If the twin on Earth could see the clock in the starship, he/she would notice it was turning much more slowly than the earthbound clock. Imagine what happens when the traveling twin returns and finds that he/she is only ten years older but the earthbound twin is 50 years older because of time dilation. The space-traveling twin will have time traveled a net 40 years into Earth's future by taking the ten-year space trip!

The second effect is more complex than time dilation, and I will simply state what it is. According to Einstein's *theory of general relativity*, objects located close to massive objects, such as the Earth, will have their clocks moving slower as compared to objects that are further away from the massive objects. This effect is due to the curvature of the space-time continuum and has been predicted and experimentally verified by the general relativity theory.

Now back to the GPS satellites that are orbiting at 14,000 kilometers per hour (km/h), while the earth is rotating at a placid 1,666 km/h. The relativistic time dilation due to the speed differences is approximately –7 microseconds per day (μs/day), while the difference due to space-time is +45 μs/day for a net satellite clock gain of 38 μs/day. While this error is nearly infinitesimal on a short-term basis, it would be very noticeable over a 24-hour time period. The total daily accumulated error would amount to a position error of 10 km or 6.2 miles (mi), essentially making GPS useless. That's why the earth ground stations constantly update and synchronize the GPS satellite atomic clocks.

NOTE **The atomic clocks within the GPS satellites are deliberately slowed prior to launch in order to counteract the relativistic effects described earlier. Ground updates are still needed to ensure that the clocks are synchronized to the desired one-nanosecond accuracy.**

The Ultimate GPS Receiver

I will be using the Ultimate GPS receiver breakout board available for about $40 from Adafruit Industries. This receiver is shown in Fig. 5-7.

This receiver meets the following comprehensive technical specifications that make it ideal for this application:

- Satellites: 22 tracking, 66 searching
- Patch antenna size: 15 mm × 15 mm × 4 mm
- Update rate: 1 to 10 hertz (Hz)
- Position accuracy: 1.8 meters
- Velocity accuracy: 0.1 meter per second (m/s)
- Warm/cold start: 34 seconds
- Acquisition sensitivity: −145 dBm
- Tracking sensitivity: −165 dBm
- Maximum altitude for PA6H: tested at 27,000 meters
- Maximum velocity: 515 m/s
- V_{IN} range: 3.0–5.5 V
- MTK3339 operating current: 25 mA tracking, 20 mA current draw during navigation
- Output: NMEA 0183, 9600 baud (Bd) default
- DGPS/WAAS/EGNOS supported
- FCC E911 compliance and AGPS support (Offline mode: EPO valid up to 14 days)

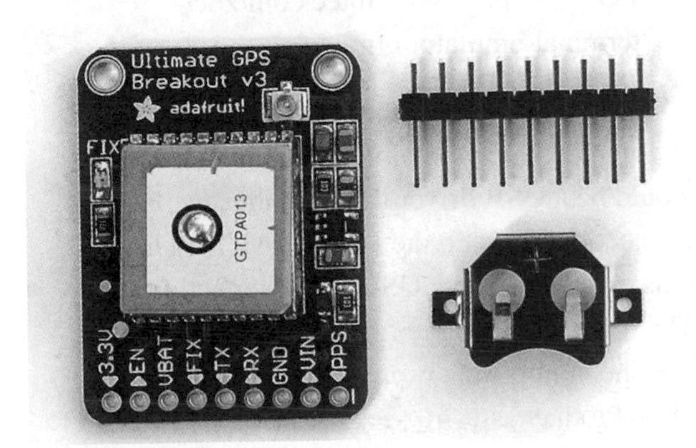

Figure 5-7 Ultimate GPS receiver breakout board.

- Up to 210 PRN channels
- Jammer detection and reduction
- Multipath detection and compensation
- Capability of an external antenna being attached
- UART for data communications. (This last feature will be discussed further in the following section, "UART Communications.")

I will neither need nor use many of these features, but they are listed here to give you an appreciation of the technical complexity and versatility of this particular GPS receiver.

There are several key specifications that are worth discussing a bit more. An acquisition sensitivity of −145 dBm means the receiver is extremely sensitive to picking up weak GPS signals. The −165 dBm tracking sensitivity means the signal, once acquired, can lose up to 90 percent of its original strength, yet remain locked in by the receiver.

Having an output operating at 9600 Bd and compliant with the *National Marine Electronics Association* (NMEA-0183) standard means the receiver generates standard GPS messages at a rate twice as fast as that of comparable receivers.

The V_{IN} range of 3- to 5.5-V matches very nicely with the RasPi 3.3-V operating voltage, thus eliminating the need for any buffer circuitry.

The 34-second start-up time is excellent and probably due in part to the extreme receiver sensitivity.

UART Communications

The *Universal Asynchronous Receiver/ Transmitter* (UART) was introduced in Chap. 1 as one of the several serial data protocols/sub-systems that the RasPi incorporates into its design. In this section, I will explore how to set up a terminal control session, also known as a console control session, by using the built-in serial UART protocol. A

Figure 5-8 UART pins.

minimum of three GPIO pins are necessary to establish a console control session. As shown in Fig. 5-8, these pins are part of the 26-pin GPIO header and have the function of *transmit* (TXD), *receive* (RXD), and *ground or common* (GND).

A simple serial communications link between a laptop computer and the RasPi will demonstrate how this type of communication functions. A terminal program running on the laptop will handle that side of the link, while the built-in, serial-protocol software will handle the RasPi side. An interconnecting cable will also be needed along with a very useful software driver. The link is set up using a USB to serial TTL cable that is connected to the GPIO header, as shown in Fig. 5-9.

Figure 5-9 USB to Serial TTL link cable.

Table 5-1 GPIO Header to USB/TTL Cable Connections

Color	GPIO header pin number	Function
Black	6	Ground
White	8	TXD
Green	10	RXD
Red	2	5 V. Caution—See warning below.

The cable has four pin connectors that are color-coded and attached to the GPIO header as detailed in Table 5-1. This cable is available from Adafruit Industries as part number 954.

CAUTION ***DO NOT CONNECT the red lead to the GPIO header*** **if you are using the "regular" micro-USB power supply. This red lead is provided to allow you to power the RasPi from the host computer's USB port; however, you cannot simultaneously power the RasPi from the micro-USB power supply and the USB-port power supply, only one or the other. My recommendation is:** ***Do not*** **use this 5-V power supply; just continue to use the regular power supply.**

There are many terminal programs available that will provide the laptop-side communications very nicely. Two recommendations are:

Tera Term—the Tera Term Project—http://ttssh2.sourceforge.jp/index.html.en

ZOC—http://www.emtec.com/zoc/terminal-emulator.html

You will need one more vital software piece to complete the communications link. This would be a USB driver that provides the logical connection between the RasPi UART/TTL pins and the laptop's USB port connection. The driver is available from the Prolific website at http://www.prolific.com.tw/US/ShowProduct.aspx?p_id=225&pcid=41, which takes you to the PL2303 Windows driver

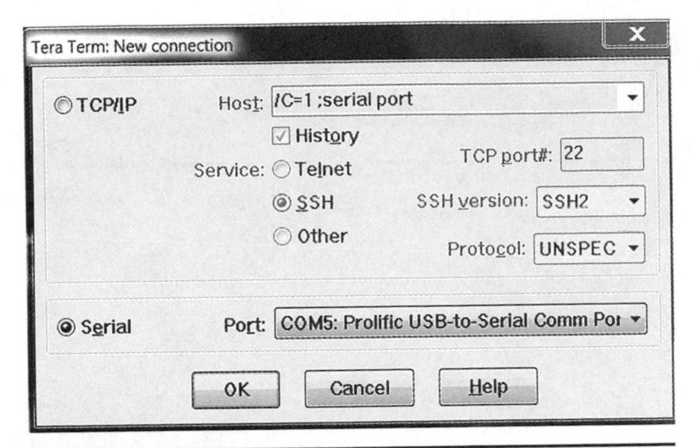

Figure 5-10 Terminal program port selection.

download page. The driver file name is PL2303_Prolific_DriverInstaller_v1.7.0.zip. This driver creates what is known as a *virtual comm port* that allows a USB port to emulate a standard RS-232 serial port. The comm port created in my case was comm5; however, yours may vary depending upon your laptop's configuration. In any case, use the new comm port in configuring the serial link, as I have shown in Fig. 5-10. In this screenshot, I was using the Tera Term program.

The port configuration settings are shown in Fig. 5-11. You click on Setup, then Serial Port to get to this screen, when using the Tera Term program.

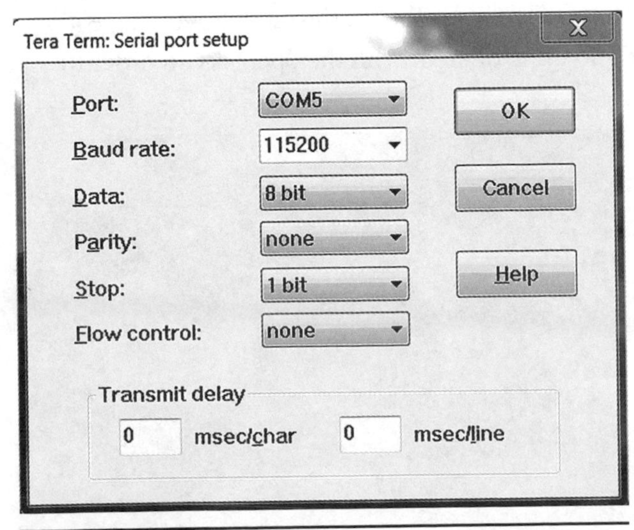

Figure 5-11 Serial port configuration settings.

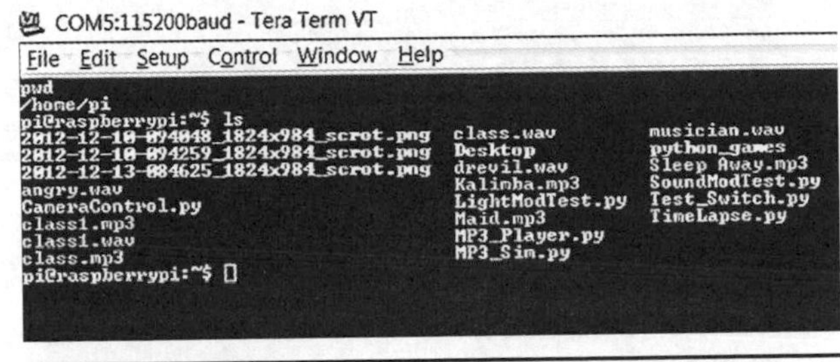

Figure 5-12 Sample screenshot from Tera Term program connected to RasPi.

The key parameters to set, independent of the terminal program that you are using, are the following:

- Baud rate—115,200
- Number of data bits—8
- Parity—None
- Stop bits—1

You should see a screen similar to the screenshot in Fig. 5-12, if you have setup everything correctly including the Prolific driver.

Figure 5-13 shows another screenshot using the ZOC terminal control program connected to the RasPi.

In a later chapter, I will show you how to connect to the RasPi over a network by using another type of terminal control program in lieu of using a direct serial cable. But for now, you have the background and familiarity to set up a serial link with a GPS receiver.

GPS Receiver UART Communication

The Ultimate GPS receiver uses a 9600-Bd UART to communicate with the controlling microprocessor in order to both receive and transmit data back and forth. The UART interface pins available on the Ultimate GPS receiver are shown in Fig. 5-14. There are only three pins that we need to use for data communication, as was discussed in the previous section.

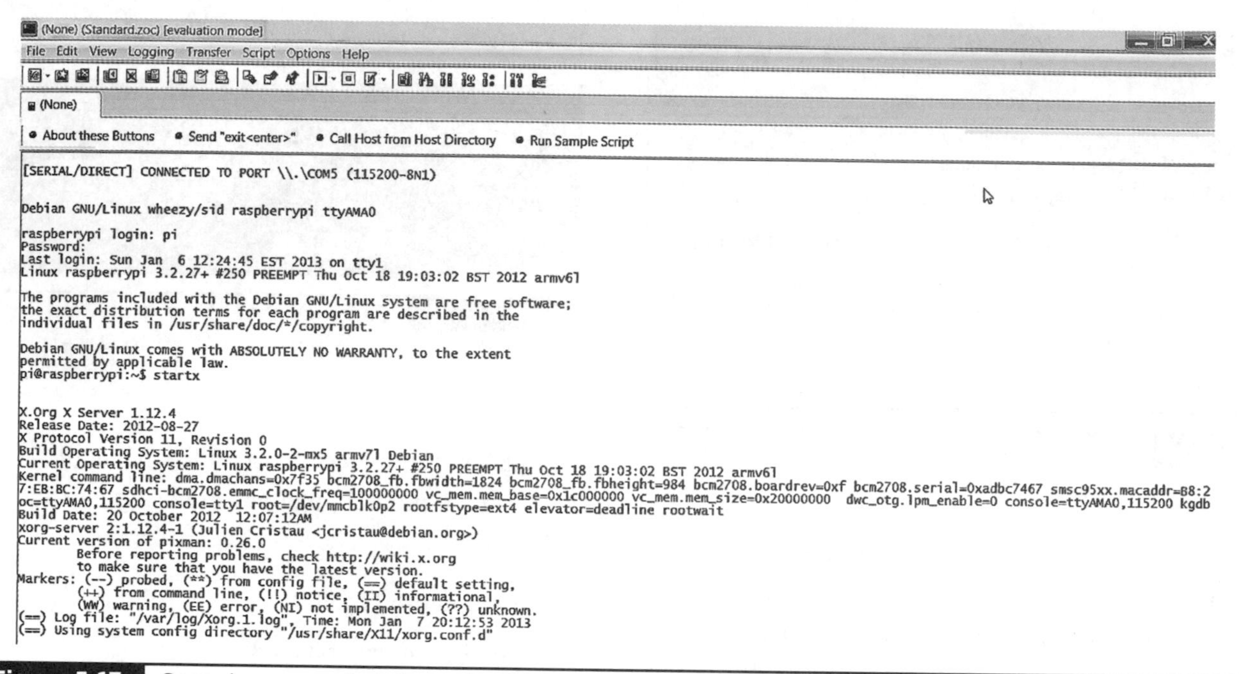

Figure 5-13 Sample screenshot from the ZOC program connected to the RasPi.

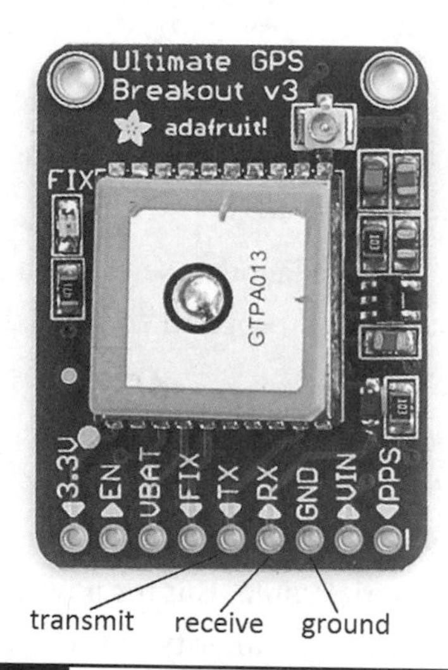

Figure 5-14 Ultimate GPS data communication interface pins.

There is no need for a separate clock signal line, since the UART protocol is designed to be "self-clocking."

Next, you need to make the connections between the Pi Cobbler UART pins and the GPS receiver pins, as shown in Fig. 5-15.

CAUTION **Ensure that the TX connector from the GPS receiver connects to the Pi Cobbler RX pin, and likewise, that the RX connector from the GPS receiver connects to the Pi Cobbler TX pin. *Do not connect* RX to RX or TX to TX even though that may seem like the logical action to take. You will not damage anything, but the data communications between the GPS receiver and the RasPi cannot be established if you make those connections.**

Initial GPS Receiver Test

It would be wise to check that the Ultimate GPS receiver is functioning as expected prior to running any code on the RasPi. Ensure that you have a good line of sight with the open sky in order to

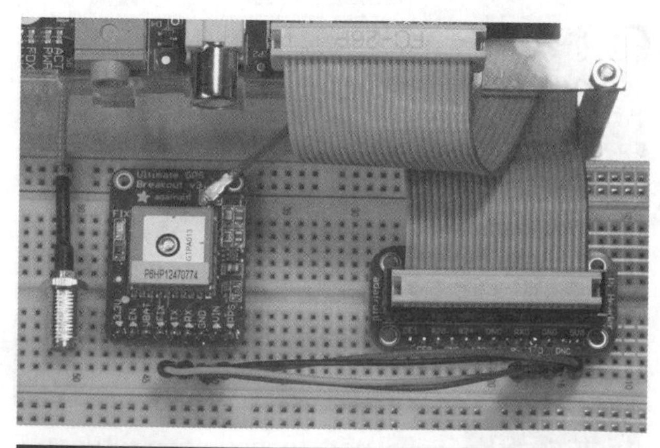

Figure 5-15 Pi Cobbler to GPS receiver connections.

receive the GPS satellite signals. I used an external GPS antenna, since my test setup was indoors without any reliable satellite reception. The antenna was purchased from Adafruit Industries, part number 960, and is well worth the modest cost because erratic or unreliable satellite reception will quickly cause this project to fail. You will also need an antenna adapter to connect the external antenna's SMA connector to the μFL connector situated on the Ultimate GPS receiver board. This adapter was also purchased from Adafruit, part number 851, and is shown in Fig. 5-15. A word of caution: Be very careful when pushing the μFL connector into the board because the inner pin appears to be quite fragile and probably would be damaged if excessive pressure were applied.

The quickest and easiest approach for a data connection is to temporarily connect the laptop to the GPS receiver with the USB/TTL cable using the connections shown in Fig. 5-16. You can parallel connect to the existing TXD and GND on the solderless breadboard without a problem.

I used the Tera Term program with the baud rate set to 9600 to match the GPS receiver output. Figure 5-17 is a screen capture of the GPS data stream showing that the GPS receiver was properly functioning and receiving good satellite signals.

The next step in confirming proper GPS operation is to disconnect the USB/TTL cable

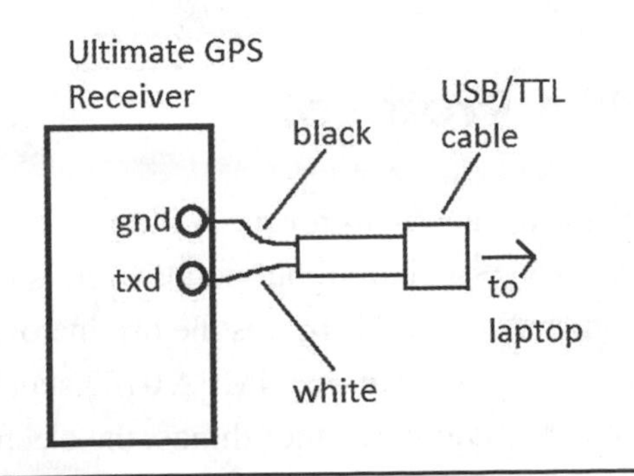

Figure 5-16 USB/TTL cable connection from GPS to laptop.

and load the RasPi with a terminal program to confirm that the RasPi can also receive the GPS data stream. I used a very nice serial terminal control program that runs in a *graphical user interface* (GUI) named *CuteCom*. To install this program, type the following into a terminal window:

```
sudo apt-get install cutecom ↵
```

NOTE I had to type "*sudo apt-get update*" prior to entering the above command because the *apt-get* repository didn't initially find the CuteCom package.

Next, you must edit the file named inittab located in /etc directory in order for CuteCom to communicate with the GPS receiver. Change the line below into a comment line.

```
TO:23:respawn:/sbin/getty —L ttyAMA0
115200 VT100
```

I used the nano editor to place the '#' symbol at the start of this line. This action is required to allow Linux to free up the serial console resource.

You can run the CuteCom program after editing inittab, by using File Manager and opening its icon located in the "other" folder in the "Applications" folder. You must enter the appropriate configuration data in the CuteCom GUI as follows:

- Device—ttyAMA0
- Baud Rate—9600
- Data Bits—8
- Stop Bits—1
- Parity—none

Figure 5-18 is a screenshot taken from the RasPi display showing the CuteCom program with the GPS data stream.

COM5:9600baud - Tera Term VT

File Edit Setup Control Window Help

```
$GPGGA,021705.000,4314.2670,N,07102.7413,W,1,4,3.74,72.4,M,-32.8,M,,*5A
$GPGSA,A,3,29,23,31,25,,,,,,,,,3.87,3.74,0.99*01
$GPGSV,2,1,05,31,64,049,39,23,29,310,35,29,27,065,40,25,10,046,30*78
$GPGSV,2,2,05,35,,,*7A
$GPRMC,021705.000,A,4314.2670,N,07102.7413,W,0.17,18.35,120113,,,A*40
$GPVTG,18.35,T,,M,0.17,N,0.31,K,A*06
$PGTOP,11,3*6F
$GPGGA,021706.000,4314.2680,N,07102.7413,W,1,4,3.74,73.3,M,-32.8,M,,*50
$GPGSA,A,3,29,23,31,25,,,,,,,,,3.86,3.74,0.99*00
$GPRMC,021706.000,A,4314.2680,N,07102.7413,W,0.14,18.35,120113,,,A*4F
$GPVTG,18.35,T,,M,0.14,N,0.26,K,A*03
$PGTOP,11,3*6F
$GPGGA,021707.000,4314.2687,N,07102.7412,W,1,4,3.74,74.0,M,-32.8,M,,*53
$GPGSA,A,3,29,23,31,25,,,,,,,,,3.87,3.74,0.99*01
$GPRMC,021707.000,A,4314.2687,N,07102.7412,W,0.21,18.35,120113,,,A*4E
$GPVTG,18.35,T,,M,0.21,N,0.38,K,A*0A
$PGTOP,11,3*6F
$GPGGA,021708.000,4314.2697,N,07102.7411,W,1,4,3.74,74.7,M,-32.8,M,,*59
$GPGSA,A,3,29,23,31,25,,,,,,,,,3.87,3.74,0.99*01
$GPRMC,021708.000,A,4314.2697,N,07102.7411,W,0.36,18.35,120113,,,A*45
$GPVTG,18.35,T,,M,0.36,N,0.66,K,A*07
$PGTOP,11,3*6F
$GPGGA,021709.000,4314.2704,N,07102.7409,W,1,4,10.08,75.4,M,-32.8,M,,*61
$GPGSA,A,3,29,23,31,25,,,,,,,,,11.72,10.08,5.97*0A
$GPRMC,021709.000,A,4314.2704,N,07102.7409,W,0.25,18.35,120113,,,A*44
$GPVTG,18.35,T,,M,0.25,N,0.47,K,A*06
$PGTOP,11,3*6F
$GPGGA,021710.000,4314.2709,N,07102.7408,W,1,4,3.74,75.6,M,-32.8,M,,*5E
$GPGSA,A,3,29,23,31,25,,,,,,,,,3.87,3.74,0.99*01
$GPGSV,2,1,05,31,64,049,39,23,29,310,35,29,27,065,40,25,10,046,30*78
$GPGSV,2,2,05,35,,,*7A
$GPRMC,021710.000,A,4314.2709,N,07102.7408,W,0.15,18.35,120113,,,A*43
$GPVTG,18.35,T,,M,0.15,N,0.28,K,A*0C
$PGTOP,11,3*6F
$GPGGA,021711.000,4314.2708,N,07102.7408,W,1,4,3.74,75.5,M,-32.8,M,,*5D
$GPGSA,A,3,29,23,31,25,,,,,,,,,3.87,3.74,0.99*01
$GPRMC,021711.000,A,4314.2708,N,07102.7408,W,0.13,18.35,120113,,,A*45
$GPVTG,18.35,T,,M,0.13,N,0.24,K,A*06
$PGTOP,11,3*6F
$GPGGA,021712.000,4314.2700,N,07102.7411,W,1,4,10.12,74.7,M,-32.8,M,,*6F
$GPGSA,A,3,29,23,31,25,,,,,,,,,11.76,10.12,5.99*0B
$GPRMC,021712.000,A,4314.2700,N,07102.7411,W,0.18,18.35,120113,,,A*4D
$GPVTG,18.35,T,,M,0.18,N,0.33,K,A*0B
$PGTOP,11,3*6F
$GPGGA,021713.000,4314.2690,N,07102.7412,W,1,4,3.74,73.9,M,-32.8,M,,*5E
$GPGSA,A,3,29,23,31,25,,,,,,,,,3.87,3.74,0.99*01
$GPRMC,021713.000,A,4314.2690,N,07102.7412,W,0.15,18.35,120113,,,A*4A
$GPVTG,18.35,T,,M,0.15,N,0.27,K,A*03
$PGTOP,11,3*6F
$GPGGA,021714.000,4314.2686,N,07102.7413,W,1,4,3.74,73.5,M,-32.8,M,,*53
$GPGSA,A,3,29,23,31,25,,,,,,,,,3.87,3.74,0.99*01
$GPRMC,021714.000,A,4314.2686,N,07102.7413,W,0.25,18.35,120113,,,A*48
$GPVTG,18.35,T,,M,0.25,N,0.46,K,A*07
$PGTOP,11,3*6F
$GPGGA,021715.000,4314.2683,N,07102.7415,W,1,4,3.74,72.9,M,-32.8,M,,*5C
$GPGSA,A,3,29,23,31,25,,,,,,,,,3.87,3.74,0.99*01
$GPGSV,2,1,05,31,64,049,40,23,29,310,35,29,26,065,40,25,10,046,30*77
$GPGSV,2,2,05,36,,,*79
$GPRMC,021715.000,A,4314.2683,N,07102.7415,W,0.22,18.35,120113,,,A*4D
$GPVTG,18.35,T,,M,0.22,N,0.41,K,A*07
$PGTOP,11,3*6F
$GPGGA,021716.000,4314.2680,N,07102.7416,W,1,4,3.74,72.4,M,-32.8,M,,*52
$GPGSA,A,3,29,23,31,25,,,,,,,,,3.87,3.74,0.99*01
$GPRMC,021716.000,A,4314.2680,N,07102.7416,W,0.18,18.35,120113,,,A*47
$GPVTG,18.35,T,,M,0.18,N,0.33,K,A*0B
```

Figure 5-17 Tera Term screen capture of GPS data stream.

Completing the steps just described confirms the proper operation of the Ultimate GPS receiver and the proper functioning of the data connection between the RasPi and the receiver.

You are almost ready to start using the GPS receiver, but first I need to discuss the NMEA protocol and the messages that are being generated from the Ultimate GPS receiver.

NMEA Protocol

NMEA is the acronym for the National Marine Electronics Association, but nobody refers to it by its formal name. NMEA is the originator and continuing sponsor of the NMEA 0183 standard, which defines, among other things, the electrical and physical standards to be used in GPS receivers.

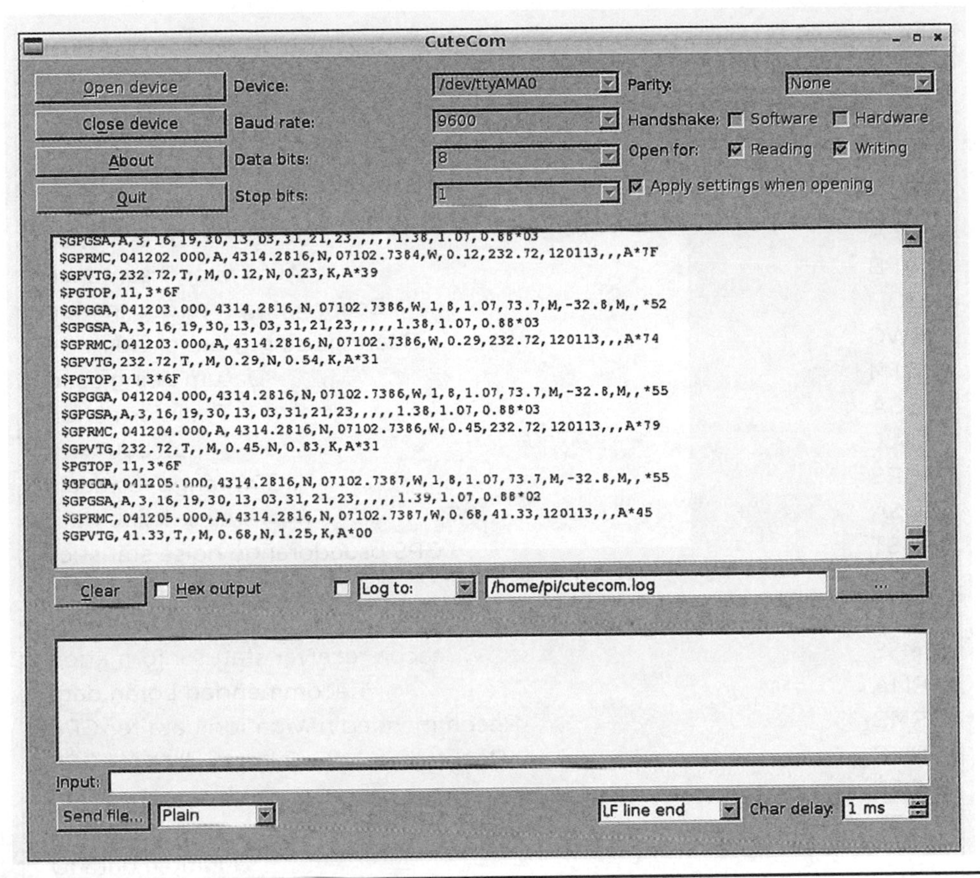

Figure 5-18 CuteCom GPS data stream.

This standard specifies a series of message types that receivers use to create messages that conform to the following rules, also known as the Application Layer Protocol Rules:

- The starting character in each message is the dollar sign.
- The next five characters are composed of the talker ID (first two characters) and the message type (last three characters).
- All data fields that follow are delimited by commas.
- Unavailable data is designated by only the delimiting comma.
- The asterisk character immediately follows the last data field, but only if a checksum is applied.
- The checksum is a two digit hexadecimal number that is calculated using a bitwise exclusive OR algorithm on all the data between the starting '$' character and the ending '*' character but including those characters.

There are a large variety of messages available in the NMEA standard; however, the following subset is applicable to the GPS environment and is shown in Table 5-2. All GPS messages start with "GP."

Latitude and Longitude Formats

The two digits immediately to the left of the decimal point are whole minutes, to the right are decimals of minutes. The remaining digits to the left of the whole minutes are whole degrees.

Table 5-2 NMEA GPS Message Types

Message Type Prefix	Meaning
AAM	Waypoint arrival alarm
ALM	Almanac data
APA	Auto pilot A sentence
APB	Auto pilot B sentence
BOD	Bearing origin to destination
BWC	Bearing using great circle route
DTM	Datum being used
GGA	Fix information
GLL	Lat/Lon data
GRS	GPS range residuals
GSA	Overall satellite data
GST	GPS psuedorange noise statistics
GSV	Detailed satellite data
MSK	Send control for a beacon receiver
MSS	Beacon receiver status information
RMA	Recommended Loran data
RMB	Recommended navigation data for GPS
RMC	Recommended minimum data for GPS
RTE	Route message
TRF	Transit fix data
STN	Multiple data ID
VBW	Dual ground/water speed
VTG	Vector track a speed over the ground
WCV	Waypoint closure velocity (velocity made good)
WPL	Waypoint location information
XTC	Cross-track error
XTE	Measured cross-track error
ZTG	Zulu (UTC) time and time to go (to destination)
ZDA	Date and time

Examples:

4224.50 is 42 degrees and 24.50 minutes or 24 minutes, 30 seconds. .50 of a minute is exactly 30 seconds.

7045.80 is 70 degrees and 45.80 minutes or 45 minutes, 48 seconds. .80 of a minute is exactly 48 seconds.

Parsed GPS Message

The following is an example of a parsed GPGLL message that illustrates how to analyze an actual data message:

```
$GPGLL,5133.80,N,14240.25,W*75
 1 2    3      4     5     6 7
```

1. GP—GPS NMEA designator
2. GLL—Lat/Lon message type
3. 5133.80—Current latitude 51 degrees, 33 minutes, 48 seconds
4. N—North/South
5. 14240.25—Current longitude 142 degrees, 40 minutes, 15 seconds
6. W—East/West
7. *75—Checksum

All GPS applications use some type of parser application to analyze data messages and extract the required information to meet system requirements. This will be discussed in the next section.

The gpsd Apps

It is time to install and run the RasPi GPS application now that the GPS receiver and serial data connection have been proved to work correctly. I will actually be using a suite of GPS tools or apps contained in a package named gpsd. Enter the following command in a Linux terminal window to download and install this suite:

```
sudo apt-get install gpsd gpsd-clients
python-gps ↵
```

gpsd is known as a daemon (pronounced day-mon) and essentially is a program that runs in the Linux OS background. The nice thing about daemons is that they go about their business without needing any attention once started. To start gpsd, type the following in a Linux terminal window:

```
gpsd /dev/ttyAMA0 ↵
```

All that's happening here is that you are telling the daemon which serial connection is being used, in this case, ttyAMA0.

Display GPS Data

The cgps app displays the GPS information, such as speed, position, altitude, etc. To see this information, type cgps in a Linux terminal window after starting the gpsd daemon.

```
cgps ↵
```

You should see the screenshot shown in Fig. 5-19 appear on the RasPi screen.

The figure shows the parsed, or analyzed, GPS data that is streaming in real time from the Ultimate GPS receiver. There is a lot of information displayed, including the following on the left-hand side:

- Date—YYYY-MM-DD
- Time—UTC accurate to 1 millisecond
- Latitude—xx.xxxxxx N or S

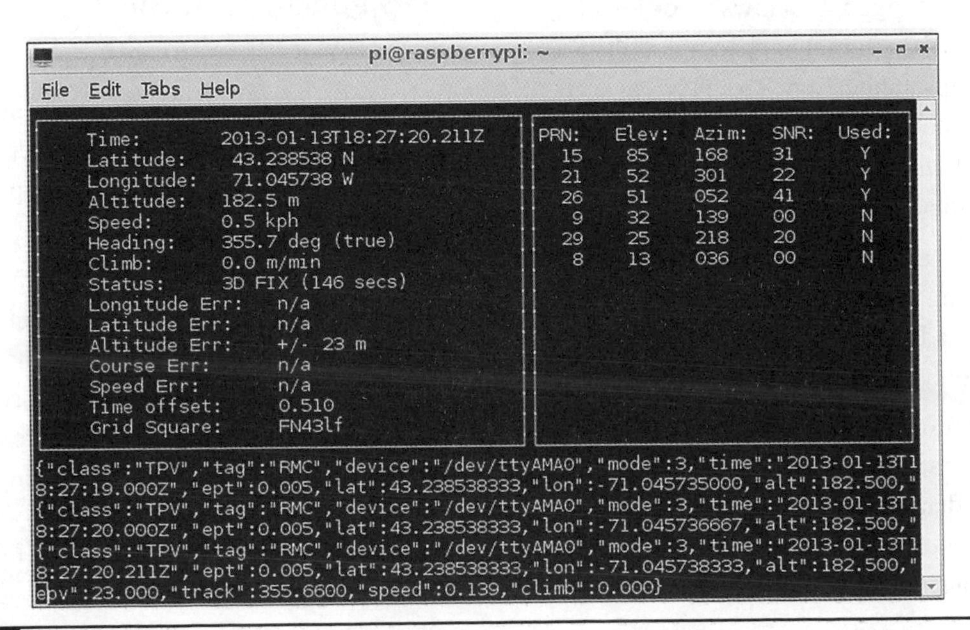

Figure 5-19 A cgps screenshot.

- Longitude—xxx.xxxxxx E or W
- Altitude—Meters (above sea level)
- Speed—Kilometers/hour (km/h or kph)
- Heading—Degrees true (not magnetic)
- Climb rate—Meters/second (applicable to aviation)
- Status—Usually 3D but maybe 2D or no fix
- Longitude error—Usually several meters
- Latitude error—Usually several meters
- Altitude error—Meters (could be 10s of meters)
- Course error—Degrees (applicable to aviation)
- Speed error—Usually a fraction of a km/h or kph
- Time offset—Difference between GPS and UTC clocks (usually ignored)
- Grid Square—The Maidenhead Locator System grid indicator

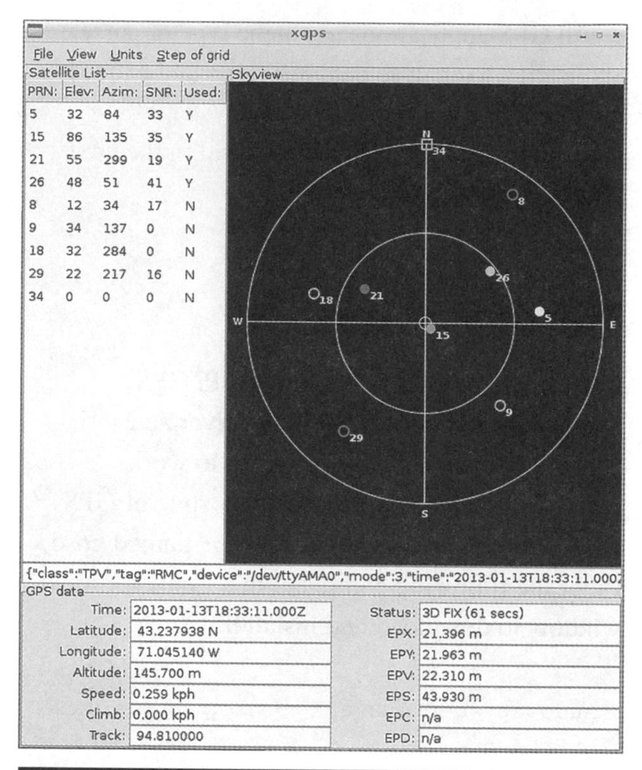

Figure 5-20 An xgps screenshot.

On the right-hand side is a real-time list of the GPS satellites in view. There were six satellites in Fig. 5-19, three of which were used in computing the location fix. In all likelihood, the satellites not used had insufficient signal strength as indicated by a low *signal to noise ratio* (SNR).

The bottom of Fig. 5-19 shows a portion of the raw GPS data stream that gpsd uses to parse data for the cgps application. The TPV tag seen repeatedly in the data stream stands for *target*, *position*, and *velocity* and is probably the most common tag in a GPS data stream.

There is also a graphical GPS display that can be run by typing xgps in a Linux terminal window. Figure 5-20 is a RasPi screenshot showing this graphical display.

The top portion of the figure is the GPS constellation that is in a real-time sky view of the receiver. The data shown is almost identical to the satellite data shown on the right-hand side of Fig. 5-19. In a similar fashion, the data shown on the bottom of Fig. 5-20 is nearly identical to the left-hand side data shown in Fig. 5-19. One difference is that the error data is abbreviated as EPX, EPY, EPV, etc., in Fig. 5-20, while it is spelled out in Fig. 5-19.

GPS Packet Monitor

The *gpsmon* app allows you to monitor the GPS data stream while providing a simplified view of the packets flowing from the receiver to the RasPi. Figure 5-21 shows a screenshot of gpsmon in operation.

When you compare Fig. 5-21 to the previous two figures you won't see a whole lot of new information, but the latitude and longitude have been converted to an easy-to-read format of degrees, minutes, and seconds. There are also some other lesser-used data shown including:

- DOP H—Horizontal dilution of precision
- DOP V—Vertical dilution of precision

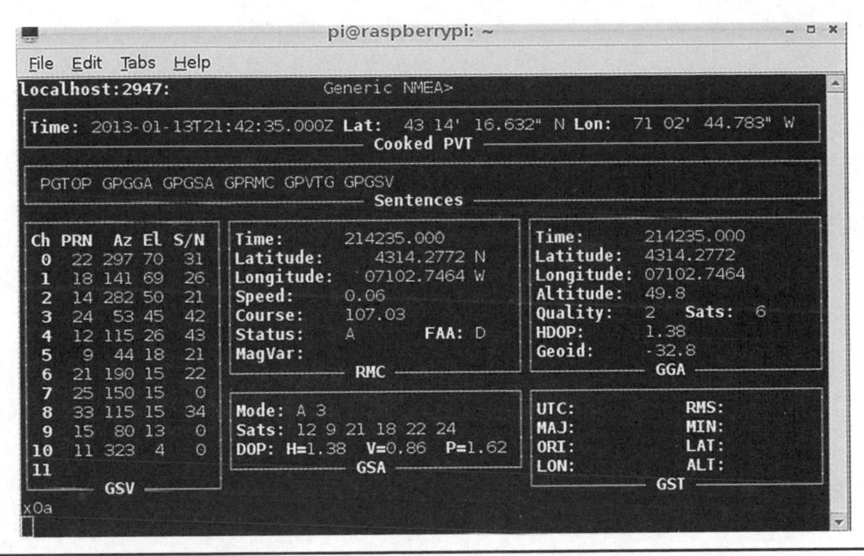

Figure 5-21 A gpsmon screenshot.

- DOP P—Total or 3D dilution of precision
- Geoid—Altitude correction due to non-spheroidal earth curvature

The *dilution of precision* (DOP) numbers are relative indicators of GPS accuracy due to the satellite–receiver configuration. Lower numbers are better, as shown in Table 5-3.

Table 5-3 Meaning of DOP Numbers

DOP	Remarks
1	Ideal, best accuracy
5–10	Moderate accuracy
>20	Poor (may be as much as 300 meters in error in a position report)

Summary

The chapter began with a brief history of the GPS system followed by a tutorial example that explained the basic underlying principles governing the system.

Next, I discussed the Ultimate GPS receiver, focusing on its excellent receiver characteristics as well as the easy serial communication link.

I discussed how to setup and test a serial console link using a USB to serial TTL cable as well as serial terminal control programs for both a Windows laptop and the RasPi. The serial comm link between the GPS receiver and the RasPi was setup, and a series of communication tests were demonstrated to verify the proper operation of all system components.

The NMEA 0183 protocol was thoroughly examined to illustrate the rich set of messages that are created by the GPS receiver. This project uses only a small subset of the data but you should be aware of what is potentially available. A parsed GPS message was also shown along with a brief explanation of how to interpret latitude and longitude data.

The remaining portion of the chapter concerned the gpsd suite of applications including cgps, xgps, and gpsmon. These apps provide an excellent set of tools to display and analyze the continuing stream of GPS data produced by the Ultimate GPS receiver.

CHAPTER 6

Earthquake Detector

Introduction

I will show you how to build a sensitive earthquake detection system using the RasPi as a controller that both processes and displays the signals generated by the detector.

Let us begin with a discussion of seismology and earthquakes, focusing on their makeup and how they are measured. *Seismology* is the term used to describe the study of earthquakes. This background section will help you understand the key concepts behind the detection-system design and what you should realistically expect from this fairly simple system.

I have also included a section on the analog-to-digital conversion that is needed to allow the RasPi to link to the earthquake, or more generally termed, seismic detector.

Some sample plots are also shown to help you understand the system outputs and to show you how you might use this system to detect imminent quakes, thereby providing you and your family with a bit more warning time in order to allow you to avoid or minimize a bad situation.

Seismology and Earthquakes

An earthquake, or more simply a quake or tremor, generally refers to any large release of energy from within the Earth's crust. There are a variety of sources that can cause this energy release, including a sudden displacement or shift in the ground, often located well below the surface level. This energy is measured in a localized area by the *Richter magnitude scale*, and globally, by the *moment magnitude scale*. The Richter scale is considered open-ended but generally ranges between 2 and 9, with level 2 quakes not being felt and level 9 quakes creating catastrophic destruction. Figure 6-1 diagrams the Richter scale plotted with quake severity designations.

The *X* axis shows the actual *Richter numbers*, which are the base-10 logarithms of values on the *Y* axis. The Y axis shows the maximum ground displacement values measured in microns, or millionths of an inch. Therefore, each increment of the Richter number represents a 10-time increase in quake severity. For example, going from a 5 to a 6 level means shifting from a moderate to a strong quake.

The most common type of quake is created by faults deep within the Earth's crust that suddenly release energy due to an almost unimaginable force build-up. This release, or ground displacement, generates two types of energy waves that rapidly transverse through the Earth. The initial energy wave is termed a *P-wave*, also known as a primary or pressure wave, and it has a velocity of approximately 1000 meters/second. P-waves, while energetic, carry far less energy than the *S-wave*, also known as a secondary or shear wave. The S-wave travels at 250 meters/second, slower

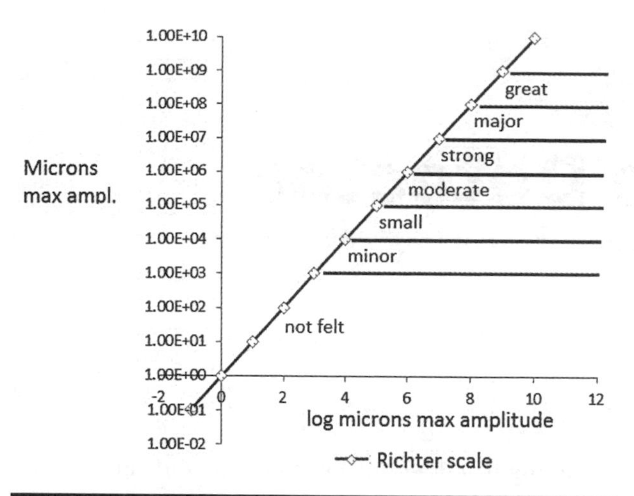

Figure 6-1 Richter scale and quake severity.

than the P-wave but containing much greater ground displacements than the P-wave. The larger amplitude S-waves cause the majority of damage to surface structures.

The seismic detector is sensitive to both P- and S-waves. The P-waves will be detected first, since they travel faster than S-waves. The difference between the time of arrival (TOA) of the two waves will depend upon the distance from the quake epicenter to the location of the detector. A distance of a few kilometers will result in only a few seconds between the waves, while that of several hundred kilometers can amount to 20 minutes or more. The RasPi can be set to generate an alarm if an initial wave greater than a threshold amount is detected. This should provide some time to take shelter in a secure and strong location, provided that the epicenter is not too close. The velocities of the P- and S-waves are very much dependent upon what makes up the ground through which the waves travel. Table 6-1 shows some typical velocities for these waves as they travel through a variety of ground types.

As you can see from the table values, both waves have higher velocities as the ground composition becomes denser. There will not be much of a time delay if only solid granite exists between the epicenter and the detector because the P-wave travels at approximately 5000 m/s, or about 3 mi/s. Interestingly, S-waves do not travel through water (as is indicated by the n/a in the water velocity column). Note that I also used average values for both P- and S-waves in the above discussion regarding expected TOAs.

Another key parameter of all seismic waves is the period of time between amplitude peaks, which can be seen in Fig. 6-2.

The inverse of a wave period is frequency and is measured in hertz (Hz). The average seismic frequency range is 0.5 to 5 Hz, which corresponds to a period range of 2 to 0.2 seconds respectively. This frequency range is quite low and has a significant impact in the design of a responsive seismic sensor, which is discussed in the next section.

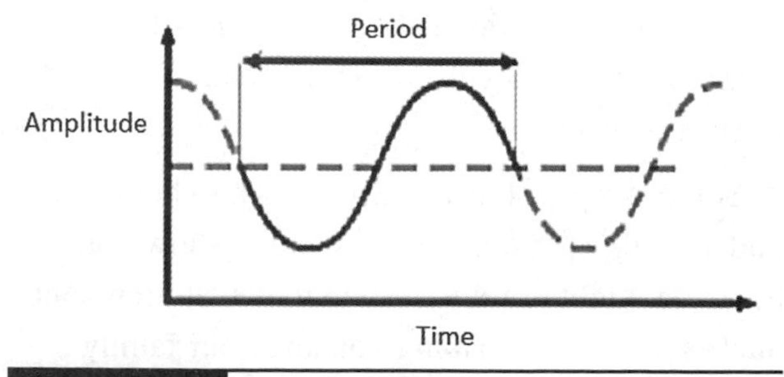

Figure 6-2 Seismic wave period.

Table 6-1 P- and S-Wave Velocities versus Ground Type

Ground type	P-wave (m/s)	S-wave (m/s)
scree, vegetal soil	300–700	100–300
saturated shale, clay	1100–2500	200–800
granite	4500–5000	2500–3300
water	1450–1500	n/a

Seismic Sensor Design

Figure 6-3 is a basic diagram illustrating the classic design for the *garden-gate seismic sensor*. It is so named because of the two pivot points that are in vertical alignment, such as would be found in a traditional, hinged garden gate. A mass is suspended between the pivots so that seismic-vibration tracings can be recorded in a horizontal direction, as shown in Fig 6-3.

Figure 6-4 is a basic diagram illustrating the classic design for the inverted-pendulum seismic sensor. A vertical mass is suspended from a horizontal-hinge assembly, and tracings are recorded in a vertical direction, as shown in the figure.

Both designs are very capable of creating mechanical paper tracings that you may have recognized from TV news reports or even disaster movies.

The modified inverted-pendulum design used in this project is very responsive to low-frequency seismic vibrations and is fairly inexpensive to construct. The sensor is a modification of a seismograph project, the "Poor Man's Seismograph" (PMS), which was described in the May 2012 edition of the *Nuts and Volts* magazine. Ron Newton created the sensor that I found very appealing because it was an extremely simple design, yet highly capable of detecting low-frequency seismic waves. The sensor produces an analog voltage representation of the seismic wave in lieu of a mechanical tracing, as is the case in the classical design.

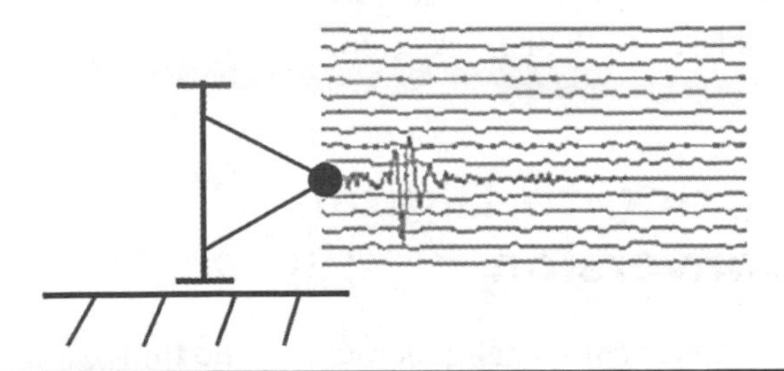

Figure 6-3 Garden-gate seismic sensor.

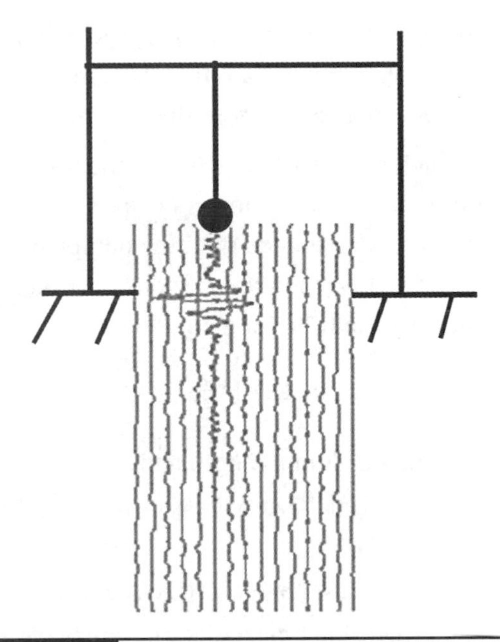

Figure 6-4 Inverted pendulum seismic sensor.

Figure 6-5 shows the actual sensor assembly that consists of a thin brass strip bolted to a thin-film-piezoelectric sensor that, in turn, is attached to a circuit board. The brass strip is 1/4 inch wide by 1/32 inch thick and 9 inches long. There is a 1/16-inch diameter hole drilled 1/8 inch on center from one end. The brass strip is bolted to the sensor using a fine-threaded #80 nut and bolt set. You should also use a drop of adhesive, such as blue Loctite, on the nut to ensure that it does not loosen.

The thin-film-piezoelectric sensor is a model LDT0-028K manufactured by Measurements Specialties Corp. It consists of a 28-µm thick, piezoelectric PVDF polymer film with screen-printed, silver-ink electrodes all laminated to a 0.125-mm polyester substrate. Two crimped contacts provide for external electrical connections.

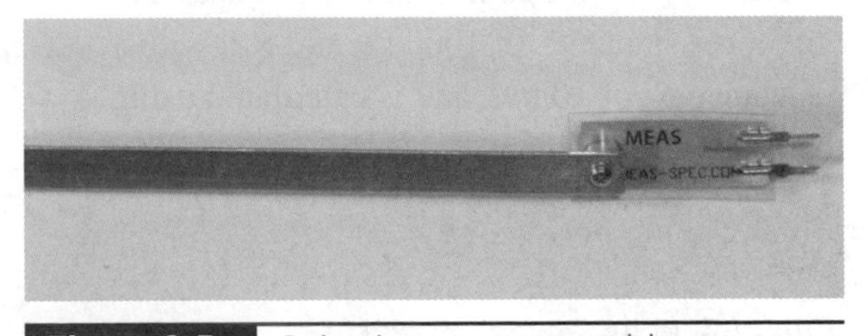

Figure 6-5 Seismic sensor assembly.

The bending forces on the piezoelectric polymer film create a very high strain that, in turn, causes high voltages to appear across the electrodes.

The 9-inch brass strip attached to the sensor adds additional mass that lowers response frequency when combined with the damping effect of the sensor. The pendulum response period t is determined by the following equation:

$$t = 2\pi\sqrt{\frac{L}{g}}$$

where L = the pendulum length and
g = the standard gravity constant, which is 9.8 m/sec^2.

In this case, L = 9.8 inches (the 9-inch brass strip plus the 0.8-inch-long piezoelectric sensor) or .249 m. Plugging in the values, we get

$$t = 2\pi\sqrt{\frac{.249\text{ m}}{9.8\text{ m/sec}^2}}$$

which yields a response period of 1 second.

$$t = 2 \times 3.14 \times .16 = 1\text{ s}$$

or a frequency (f) of 1 Hz. (Recall that the frequency is the inverse of the response period, or wave period.) This value is ideal for seismic wave detection.

The piezoelectric film sensor does need some amplification to detect weak seismic signals. Figure 6-6 shows a portion of the PMS circuit, the *electronic amplifier* (amp), that increases the signal to useful levels. The IC1 shown in Fig. 6-6 is a Microchip model MP601 *operational amplifier* (op amp) that operates from a single power supply of 3 V and is configured as a noninverting amplifier. The 11X gain is fixed by resistors R1, with a resistance R_1 of 100 kΩ, and R2, with a resistance R_2 of 10 kΩ, and is calculated using the following equation:

$$\text{Gain} = (R_1 + R_2)/R_2$$

Plugging in the values, we get

$$11 = (100\text{ k}\Omega + 10\text{ k}\Omega)\ /\ 10\text{ k}\Omega$$

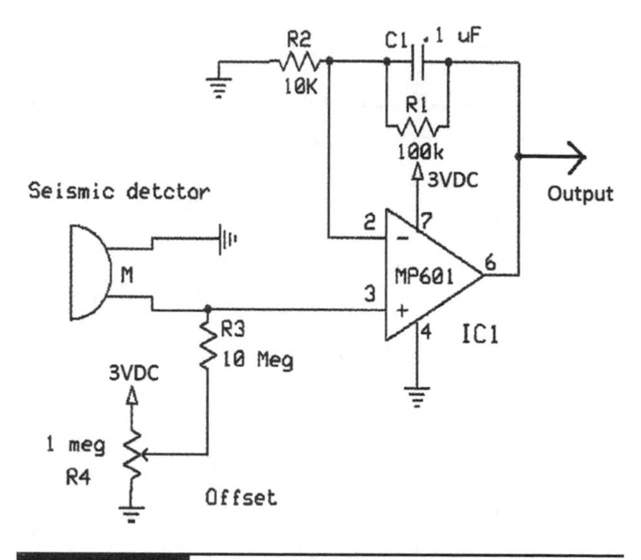

Figure 6-6 Seismic amplifier circuit.

The *potentiometer*, or variable resistor, R4 in combination with resistor R3 sets an adjustable DC voltage offset so that the full range of the amplitude swing from the seismic sensor is input to the op amp without clipping or distortion. The op amp output from pin 6 goes to both the RasPi processing circuit and the PMS microprocessor.

The original PMS uses a Microchip PIC24HJ64GA002 microprocessor to convert and store detected seismic waves onto an SD card, which is later processed on a PC to display the waves. My project uses the RasPi to process the seismic waves in real time, thus making the information immediately available. Unfortunately, the RasPi does not contain any built-in *analog-to-digital converters* (ADCs), while the PIC microprocessor does have this feature. An external ADC chip must be used to convert the analog seismic signal to the equivalent digital format, as discussed in the next section.

Analog-to-Digital Conversion

The electrical signal generated by the thin-film piezoelectric sensor is classified as analog because it is a continuous voltage representation of the mechanical vibrations affecting the sensor. The RasPi, meanwhile, can process only digital signals that have just two voltage values, 0 and 3.3 V. Therefore, an ADC process must be used to convert the seismic signal to a suitable digital representation. The first step in the ADC process is to sample the analog signal in order to capture a voltage level that will be converted to a number reflecting the sampled voltage level.

Sample rate is the term used to indicate how often the analog signal level is captured. The *Nyquist rate* is a very specific constraint that determines the real-time sample rate. This value may be expressed in samples per second (sps) or sampling frequency (*fs*) and is determined as follows:

$$fs = 2 \times fmax$$

where *fmax* is the highest-frequency component present within the signal.

In our case, the seismic signal *fmax* is 5 Hz, as discussed above; therefore, *fs* is a minimum of 10 Hz or 10 sps. This is a very low value and provides a lot of time between samples to do the necessary digital signal processing.

The second step in the ADC process is to convert the sample voltage to a digital number. This can happen in a variety of ways, all of which are dependent upon the type of ADC chip that is used. I used a Microchip model MCP3008 that is described in the Microchip datasheet as a 10-bit, SAR ADC with SPI data output. Translated, this means that the MCP3008 uses a *successive approximation register* (SAR) technique to create a 10-bit digital result that, in turn, is outputted in a serial data stream using the SPI protocol. Figure 6-7 and a short discussion will help clarify this description.

The analog signal is first selected from one of eight channels that may be connected to the input channel multiplexer. Using one channel at a time is called operating in a *single-ended mode*. The MCP3008 channels can be paired to operate in a *differential mode*, if desired. A single configuration bit named SGL/DIFF selects single-ended or differential operating modes. Single-ended is the mode used in this project.

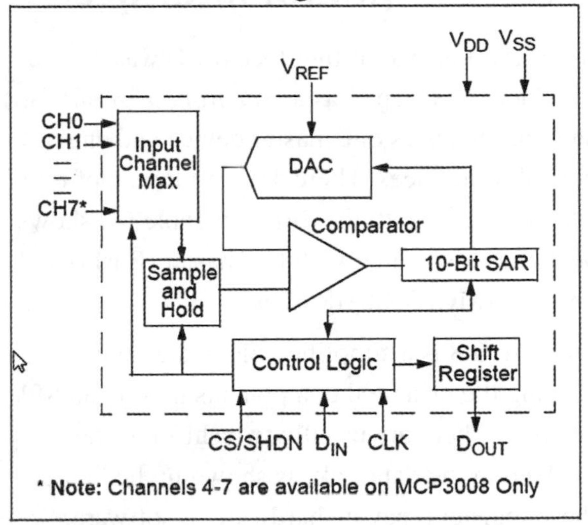

Figure 6-7 MCP3008 functional block diagram.

The selected multiplexer channel is then routed to a *sample-and-hold* circuit that is one of the two inputs to a comparator. The other input is from a *digital-to-analog converter* (DAC) that receives its input from a 10-bit SAR. Basically, the SAR starts at 0 and rapidly increments to a maximum of 1023, which is the largest number that can be represented with 10 bits. Each increment increases the voltage appearing at the DAC's comparator input. The comparator will trigger when the DAC's voltage precisely equals the sampled voltage, and this will stop the SAR from incrementing. The digital number that exists on the SAR at the moment the comparator "trips" is the ADC value. This number is then outputted, one bit at a time through the SPI circuit discussed below. All this takes place between sample intervals. The actual voltage represented by the ADC value is a function of the reference voltage V_{REF} connected to the MCP3008. In our case, V_{REF} is 3.3 V; therefore, each bit represents 3.3/1024 or approximately 3.223 millivolts. For example, an ADC value of 500 would represent an actual voltage of 1.612 V, which was computed by multiplying .003223 by 500.

Serial Peripheral Interface

The Serial Peripheral Interface (SPI) was introduced in Chap. 1 as a *synchronous serial* (SS) data link that uses one master device and one or more slave devices. There are a minimum of four data lines used with the SPI, and Table 6-2 shows the names associated with the master (RasPi) and the slave (MCP3008) devices.

Figure 6-8 is a simplified block diagram showing the principal components used in an SPI data link. There are usually two shift registers involved in the data link, as shown in the figure. These registers may be hardware or software, depending upon the devices involved. The RasPi implements its shift register in software, while the MCP3008 has a hardware shift register. In either case, the two shift registers form what is known as an inter-chip circular buffer arrangement, which is the heart of the SPI.

Data communication is initiated by the master, which begins by selecting the required slave. The RasPi selects the MCP3008 by bringing the SS line to a LOW state or 0 V. During each clock cycle, the master sends a bit to the slave that reads it from the MOSI line. Concurrently, the slave sends a bit to the master, which reads it from the MISO line. This operation is known as *full-duplex communication*, i.e., simultaneous reading and writing between master and slave. Figure 6-9 shows the Master-Slave connection between the RasPi and the MCP3008.

MCP3008

CH0 1 16 V_{DD}
CH1 2 15 V_{REF}
CH2 3 14 AGND
CH3 4 13 CLK
CH4 5 12 D_{OUT}
CH5 6 11 D_{IN}
CH6 7 10 $\overline{CS}$/SHDN
CH7 8 9 DGND

Figure 6-8 SPI simplified block diagram.

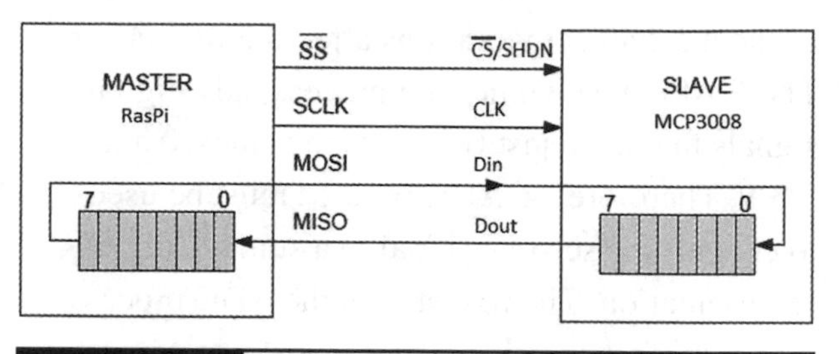

Figure 6-9 Test circuit SPI master/slave connections.

The clock frequency used is dependent primarily upon the slave's response speed. The MCP3008 can easily handle bit rates up to 3.6 MHz if powered at 5 V. Since we are using 3.3 V, the maximum rate is a bit less at approximately 2 MHz. This is still very quick and will process the RasPi input without losing any data.

The first clock pulse received by the MCP3008, with its CS pin held LOW and D_{IN} pin HIGH, constitutes the start bit. The SGL/DIFF bit follows next and then three bits that represent the selected channel(s). After these five bits have been received, the MCP3008 will sample the analog voltage during the next clock cycle.

The MCP3008 then outputs what is known as a low null bit that is disregarded by the RasPi. The following 10 bits, each sent on a clock cycle, are the ADC value with the *most significant bit* (MSB) sent first, down to the *least significant bit* (LSB), which is sent last. The RasPi will then put the MCP3008 CS pin HIGH ending the ADC process.

Table 6-2 SPI Data Line Descriptions

Master Device—RasP	Slave Device—MCP3008	Remarks
SCLK	CLK	Clock
MOSI	D_{IN}	Master Out Slave In
MISO	D_{OUT}	Master In Slave Out
CS/SHDN	SS	Slave Select

Connecting and Testing the MCP3008 with the RasPi

The MCP3008 is connected to the RasPi using the Pi Cobbler prototype tool along with a solderless breadboard. Figure 6-10 is the schematic for this circuit. The physical setup is shown in Fig. 6-11.

There is a temporary test setup on the left side of the breadboard consisting of a potentiometer connected between 3.3 V and ground. The ADC channel is connected to the tap, allowing a variable voltage that can be used in the test.

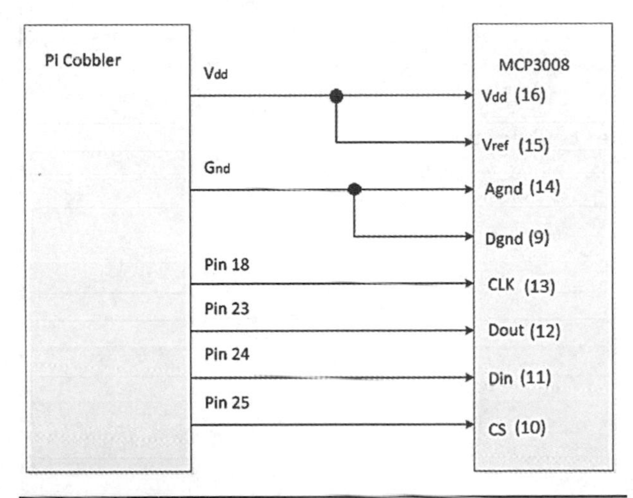

Figure 6-10 RasPi and MCP3008 connection schematic.

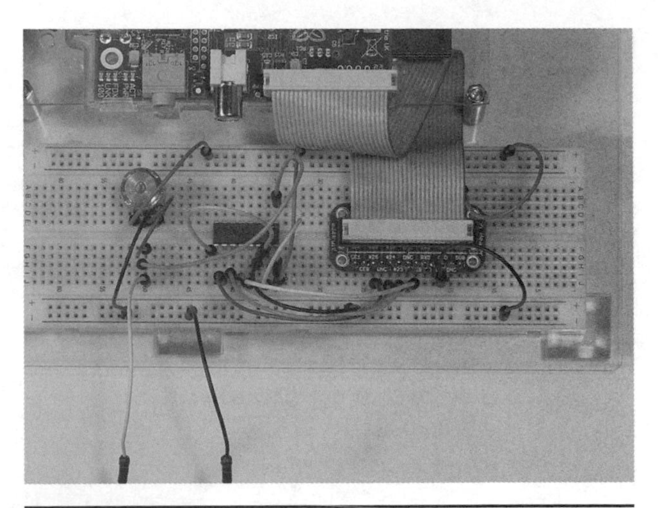

Figure 6-11 RasPi and MCP3008 test setup.

The test software that produces a continuous stream of ADC values is shown below It is also available on the book's companion website, www.mhprofessional.com/raspi, as Test_ADC.py. The code follows the ADC configuration and the SPI protocols, as discussed above. The code is based upon the sample code available from the Learn

Test_ADC.py

```
import time
import os
import RPi.GPIO as GPIO
GPIO.setmode(GPIO.BCM)
DEBUG = 1
#define a function to read the MCP3008 ADC value.
def adc(chan, clock, mosi, miso, cs):
    if((chan < 0) or (chan > 7)):
        return -1
    GPIO.output(cs, True)
    GPIO.output(clock, False)
    GPIO.output(cs, False)
    cmd = chan
    cmd |= 0x18
    cmd <<= 3
    for i in range(5):
        if(cmd & 0x80):
            GPIO.output(mosi, True)
```

```
        else:
            GPIO.output(mosi, False)
        cmd <<= 1
        GPIO.output(clock, True)
        GPIO.output(clock, False)
    result = 0
    for i in range(12):
        GPIO.output(clock, True)
        GPIO.output(clock, False)
        result <<= 1
        if(GPIO.input(miso)):
            result |= 0x1
    GPIO.output(cs, True)
    result >>= 1
    return result
#These pin definitions are set to work with the test circuit
SPICLK = 18
SPIMISO = 23
SPIMOSI = 24
SPICS = 25
GPIO.setup(SPIMOSI, GPIO.OUT)
GPIO.setup(SPIMISO, GPIO.IN)
GPIO.setup(SPICLK, GPIO.OUT)
GPIO.setup(SPICS, GPIO.OUT)
channel = 0
while True:
    #read an ADC value.
    adc_value = adc(channel, SPICLK, SPIMOSI, SPIMISO, SPICS)
    if DEBUG:
        print "value = ", adc_value
    #wait a second and repeat.
    time.sleep(1)
```

.Adafruit.com website in their discussion of the MCP3008.

Figure 6-12 presents a screenshot of a portion of the program output with the analog voltage adjusted to a 500 *value*, or *count* as it is sometimes called. An actual voltage of 1.629 V was measured using an uncalibrated *voltage output meter* (VOM). This is quite close to the computed value of 1.612. The difference was primarily due to a slightly higher supply voltage measured at 3.32 V that, when factored into the computation, yields a 1.621 V or only a .008 V difference. This is equivalent to about 3 counts or +0.3%, which is quite normal for this type of ADC.

The sharp-eyed reader may have noticed that the program does not rely on the built-in SPI functionality but instead implements a "bit-banged" interface that was discussed in Chap. 1 page 13. This approach was taken because the Linux version used in this project did not directly implement the SPI protocol. It makes no functional difference which approach is taken other than that four GPIO pins are dedicated in the "bit-banged" version and are not otherwise available for other uses.

Connecting the Seismic Detector

It is very simple to connect the PMS detector to the MCP3008 circuit. First disconnect and remove the

Figure 6-12 Test_ADC.py program output.

test circuit that was used in the above section. Then connect the PMS op-amp output and ground to the MCP3008 channel 0 and ground, respectively. Figure 6-13 shows the PMS connected to the MCP3008 circuit that, in turn, is connected to the RasPi through the Pi Cobbler.

The software also requires a slight modification because we are done with the initial testing and now need to collect some actual data from the seismic sensor. The code segment in the code listing titled "Test_ADC.py Segment (modification to allow the collection of seismic data)" on page 82 replaces all the code starting at the line shown below.

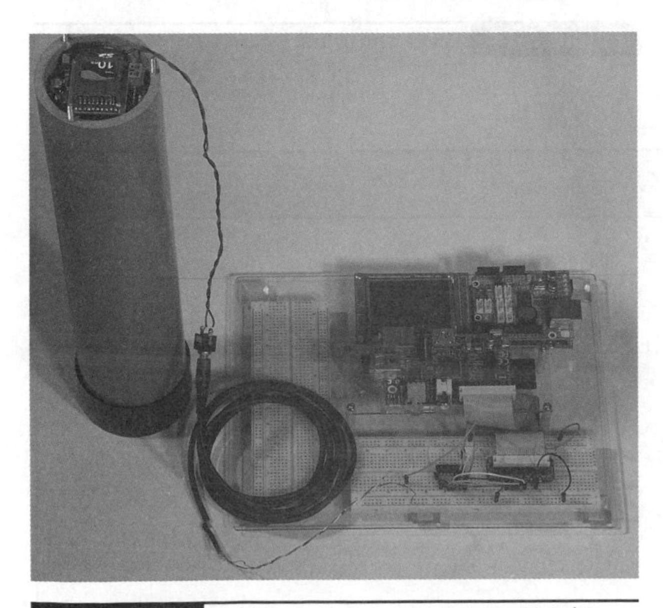

Figure 6-13 PMS connected to MCP3008/Pi Cobbler and RasPi..

```
channel = 0.
```

The program with the new code inserted is available on the book's website as Test_File_ADC .py. This new program writes 1200 `adc_values` to a file named "myData," which is located in the pi working directory, if you have not changed that location. Each value has a new-line character appended to enable the data file to be easily imported into Excel for further analysis.

Python supports two ways to store data in a file by using either string or binary formats. The approach taken in this program was to store the data as a sequence of strings, since we knew beforehand that the data would be input to an Excel spreadsheet for further analysis. Storing data as strings consumes a lot of file space as compared to storing the data in a binary format. The following code line shows how to store the data in a binary format:

```
file.write(adc_value, "wb")
```

where wb is short for write binary. Of course, you would need to read in the data using the complementary "rb", short for read binary. Not only that, but you need to restructure the program to keep track of the number of bytes read to ensure that you don't mix up the byte sequence and start reading "garbage." You also will not need the line:

```
adcStr = str(adc_value) + "\n"
```

because the binary data is written in a contiguous byte stream that you, as the programmer, must decipher. This is another reason to use strings—it is much easier to debug the file contents.

This program takes two minutes to run, as the time between samples has been reset to .1 second. Recall that this sample frequency is 10 Hz, which was discussed earlier as an optimal value for this application. Also note that "DEBUG = 0" is set, or you will get 1200 values displayed on your screen.

You should run Test_File_ADC.py while moderately shaking the PMS to create pseudo-seismic activity. Do not be so vigorous as to have the brass pendulum strike the PVC pipe wall. The sensor is quite sensitive, so only gentle force is needed. We will be ready for some data analysis once the file is generated.

Seismic Data Analysis

Copy and paste the contents of the myData file into an Excel spreadsheet. There should be exactly 1200 values in a single column. Each value represents the digitized value of a sample from the PMS detector. The samples were taken at an interval of .1 second; hence, the entire record spans two minutes. During this time, the PMS was subjected to some shaking, which provided the data for analysis. My Excel file is available on the book's companion website as Seismic.xlsx.

Figure 6-14 is a graph of an interesting portion of the data set. I selected it by visually scanning a graph of the entire record. This portion covers the first 200 samples. I next wanted to zoom in on, and

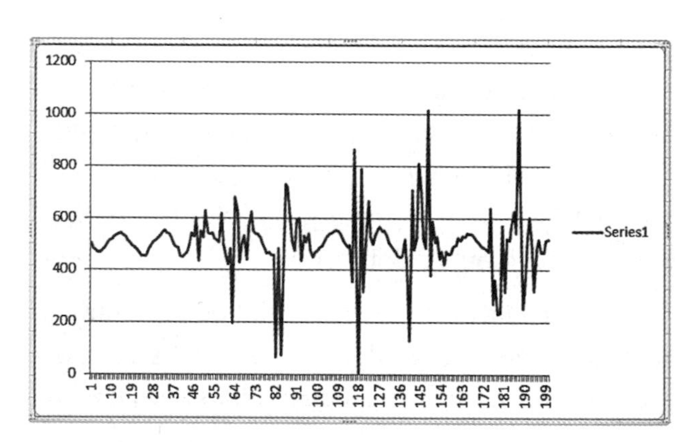

Figure 6-14 200 samples.

Test_ADC.py Segment (modification to allow collection of actual data)

```
file = open("myData", 'w')
for i in range(1200):
    adc_value = adc(channel, SPICLK, SPIMOSI, SPIMISO, SPICS)
    #str() converts the numeric adc_value into a string. The "\n" is the newline char.
    adcStr = str(adc_value) + "\n"
    file.write(adcStr)
    if DEBUG:
        print "value = ", adc_value
    #sample 10 times per second.
    time.sleep(.1)
file.close()
    print("All done")
```

examine in finer detail, a portion of this graph that covered *X*-axis numbers 139 to 168. Figure 6-15 shows this detail.

Four peaks, located at *X*-axis numbers 8, 11, 14, and 18 in Fig. 6-15, are clearly visible. These peaks are separated in time by .3, .3, and .4 second, as each *X*-axis number represents a .1 second interval. The average of the peak separation is .33 second, or equivalently, 3 Hz. This frequency is the apparent resonant frequency of this PMS detector. It falls well within the acceptable frequency span discussed earlier and is very suitable for seismic detection purposes.

The other analysis aspect that must be considered is the seismic magnitude, or *Y* axis. The highest value recorded within the two minute data record was slightly over 1000. Recall that the 10-bit MCP3008 can measure up to 1023, using a 3.3-V reference. Again, the maximum is OK, but we do have to be aware of the minimum value, or else the detector range will "bottom out".

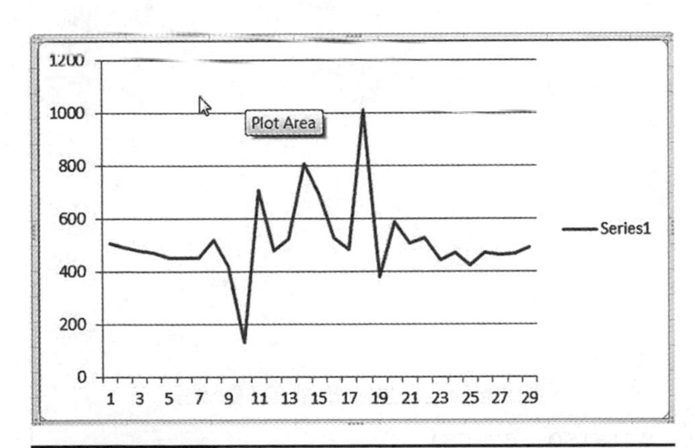

Figure 6-15 30 samples.

Examining the graphs indicates that the equilibrium level is around 500 counts. This is ideal in that an equal range, both high and low, is available to capture the sampled data. You may have to adjust your PMS if the equilibrium is not at this ideal level. That adjustment is easily done by turning the PMS offset potentiometer R4 until the displayed level is approximately 500. Refer to Fig. 6-6 to identify this control. Also, you can run the Test_ADC.py program to make this adjustment.

The final step in the analysis is to determine a valid warning or threshold level. This is somewhat subjective in that you need to set the level high enough to forestall false alerts, yet low enough to accurately capture real seismic events. Examining the data set again reveals that a level of 720 would probably suffice. It is sufficiently high to avoid random vibrations that may be caused by heavy trucks rolling nearby; yet it is low enough to capture real activity. In any case, the value can easily be reprogrammed, if needed.

Operational System

The primary requirement for an operational system is to create a warning if the detected seismic activity exceeds a preset or threshold level. I determined that a 720 level would be a good starting point. The following program, Seismic_Monitor.py, will light an LED if a level of 720 or more is detected. The LED will remain on until reset by restarting the program. Seismic_Monitor.py is available on the book's website.

Seismic_Monitor.py

```
import time
import os
import RPi.GPIO as GPIO
GPIO.setmode(GPIO.BCM)
def adc(chan, clock, mosi, miso, cs):
    if((chan < 0) or (chan > 7)):
        return -1
```

```
    GPIO.output(cs, True)
    GPIO.output(clock, False)
    GPIO.output(cs, False)
    cmd = chan
    cmd |= 0x18
    cmd <<= 3
    for i in range(5):
        if(cmd & 0x80):
            GPIO.output(mosi, True)
        else:
            GPIO.output(mosi, False)
        cmd <<= 1
        GPIO.output(clock, True)
        GPIO.output(clock, False)
    result = 0
    for i in range(12):
        GPIO.output(clock, True)
        GPIO.output(clock, False)
        result <<= 1
        if(GPIO.input(miso)):
            result |= 0x1
    GPIO.output(cs, True)
    result >>= 1
    return result
SPICLK = 18
SPIMISO = 23
SPIMOSI = 24
SPICS = 25
LED = 27 #This is pin 21 for rev 1 boards.
GPIO.setup(SPIMOSI, GPIO.OUT)
GPIO.setup(SPIMISO, GPIO.IN)
GPIO.setup(SPICLK, GPIO.OUT)
GPIO.setup(SPICS, GPIO.OUT)
GPIO.setup(LED, GPIO.OUT)
channel = 0
while True:
    adc_value = adc(channel, SPICLK, SPIMOSI, SPIMISO, SPICS)
    if(adc_value >= 720):
        GPIO.output(LED, GPIO.HIGH)
    time.sleep(.1)
```

The system functioned as expected with the LED being activated by a slight shake applied to the PMS detector. Use the nano editor to change the line

```
if(adc_value >= 720):
```

to any value that you want to set as an appropriate threshold level.

Summary

This chapter began with a brief review of seismology and how earthquakes are generated. It was pointed out that quakes produce both P- and S-waves with the P-waves being precursors to the destructive S-waves.

Next, I covered the basics of seismic sensor designs. The inverted pendulum was selected for this project because it is both sensitive and very inexpensive.

The analog-to-digital conversion process was discussed, as it is required in order to convert the analog seismic signals to a digital format suitable for processing in the RasPi.

A thorough discussion of the Serial Peripheral Interface (SPI) followed. This is the interface through which the data flows between the MCP3008 ADC chip and the RasPi.

Next the ADC chip was set up to function with the RasPi, using the Pi Cobbler prototyping tool. Also shown was a test program using "bit-banging" to control the ADC chip and implement the SPI protocol.

A seismic detector based on a kit called the "Poor Man's Seismograph" (PMS) was connected and tested with a program that collected data samples and stored them in a file on the RasPi file system. This data file was then analyzed by using MS Excel to determine a realistic alarm level.

Finally, the project concluded with the demonstration of an operational system that lit an LED when the seismic level went above the trip level.

CHAPTER 7

Home Automation

Introduction

In this chapter, I will cover how to effectively use the RasPi in a home automation system. But how do you define a home automation system? The answer is after you examine the needs and requirements of the homeowners, you design an automated system that best meets those needs. Hence, one size or type of automated system does not serve all; instead you start with a flexible base system that can be easily tailored to fit a specific situation.

The real purpose of a home automation system is to simplify a homeowner's life. It accomplishes this goal by removing repetitive and tedious tasks and relegating them to a hardware and software system that never forgets, and accurately and consistently carries out the programmed tasks. Some of the tasks that a home automation system could handle are shown in Table 7-1.

There are numerous base-system technologies available that have been developed over recent years to implement home automation. The most popular are listed in Table 7-2 with a brief description.

I selected the Z-Wave technology to use in this project because it is wireless, very robust, based on standards, and has components that are easily purchased and available from reliable manufacturers, including GE, Black & Decker, Schlage, ADT, and Draper. Anyone can purchase a Z-wave remote control and several nodes, and have an automation system functioning in a matter of minutes; the simplicity of setting up an automation system is what attracted me to this technology. This project replaces the remote with the RasPi and an attached Z-wave dongle. Of course, the RasPi is programmed to suit the homeowner's needs.

There is no hardware construction involved in this project (the only physical task is to plug in a USB dongle), but that does not mean it is easy. A lot of software setup and configuration is involved in making the project run successfully. Of course, that's part of the fun of working with the RasPi and Linux—many functions can be done in software versus building hardware peripherals.

Z-Wave Mesh Network

Each Z-Wave component or node contains a low-power *radio-frequency* (RF) transceiver. This radio operates in the *Industrial, Scientific, and Medical* (ISM) band at a frequency of 908.42 MHz in the United States and 860 MHz in Europe. These frequencies are far removed from other home wireless devices that typically operate at 2.4 GHz. It is thus far less susceptible to causing, or being subjected to, interference. The outdoor range, or line of sight, is approximately 100 m or 300 ft. The indoor range is about 30 m or 90 ft, due mainly to attenuation caused by the interior walls. Range extension is easily accomplished by simply adding additional nodes into the home.

Table 7-1 Home Automation Tasks

Category	Description
Lighting	Activate lights based upon a • schedule, • motion sensor, or • remote control. Dim lights per schedule. Control lighting IAW time of day or ambient light conditions.
Shades/privacy	Actuate motor controlled shades per • schedule, • room temperature, and/or • privacy concerns.
Home security	Activate door locks based on sensor input. Initiate alarm notification based upon sensor type and location to include • fire/smoke, • water-pipe breaks, sump-pump operation, • gas, and/or • glass breakage.
Audio/visual	Distribute A/V to specific rooms per schedule. Maintain a room intercom. Send control signals to a multimedia system.
Energy management	Measure energy consumption by • whole house, • individual appliances, and/or • HVAC.
Other	Control other domestic systems including • Irrigation system with moisture sensors, • Garage door operation, • Appliance operation, such as a coffeemaker, • Pool maintenance including preset backwash filter operation, • Spa operation including beginning a warm-up at a preset time, and/or • Pet feeding system.

Since each node is a transceiver, it both receives and transmits digital messages, acting as a digital repeater or "digipeater." Digipeating creates a signal hop. There is, however, one major proviso: only four hops are allowed. After that, the communications protocol automatically terminates the signal in a process known as hop kill. This is done presumably to limit range and probably to remain in legal compliance with governmental regulations regarding low power, unlicensed operations.

Basics of Z-Wave Network Operation

The Z-Wave network, from its very beginning, was designed to be compliant with the ISO seven-layer network model. As such, its inherent design is based upon proven computer network concepts that are robust, efficient, and well understood by most system designers. Figure 7-1 shows the Z-Wave logical network stack with the corresponding ISO layer

Table 7-2 Popular Home Automation Technologies

Name	Description
X-10	X-10 is the oldest of all these technologies that use the powerline for communications. It is often limited by phase differences, noise, and line distances. It is somewhat difficult to configure and is subject to erratic performance.
UPB	UPB is short for *Universal Powerline Bus* and was created to overcome many of the X-10 problems. It is also not compatible with the X-10 technology, thus requiring a controller that can talk to both UPB and X-10 if you are using components from each system.
INSTEON	The first home automation technology to use both powerline and wireless communication. It is fairly simple to configure and run.
Z-Wave	A totally wireless technology that is very simple to configure and run. Individual nodes also extend the total network range by the use of a digital repeater mode. It is also standards based.
ZigBee	This recent addition to home automation uses wireless. It is also used in low-power control area networks. It is somewhat complicated to configure and runs into problems when components from different manufacturers are being used.

number. All subsequent network software developed for the Z-Wave network follows this model.

Data sent through the Z-Wave network is in packets similar to the Ethernet format. Figure 7-2 shows how these packets are initially constituted at Layer 2 and subsequently modified at higher layers, as needed to suit the real-time network communication need.

A basic packet is created at the Data Link Layer 2 that starts with a synchronization preamble followed by a *Start of Frame* (SoF) byte. Next comes the payload, which can be up to 64 data bytes and is terminated by an *End of Frame* (EoF) byte.

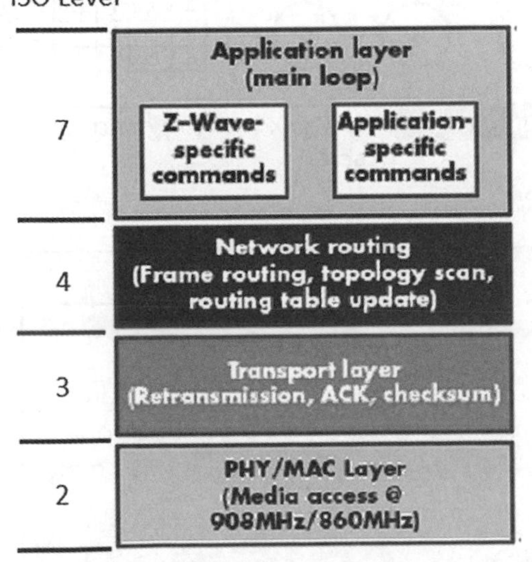

Figure 7-1 Z-Wave and ISO network layers.

The Transport Layer 3 takes the basic packet and adds additional bytes, depending upon what is required in the communication process. Z-Wave is a connection-type network, similar to Ethernet, which has a very robust way of ensuring that packets get where they need to go. Layer 3 uses *Acknowledgement* (ACK) and *Negative Acknowledgement* (NACK) packets to maintain solid connections. A receiver node will send back an ACK packet to the sender if it successfully receives a data packet; otherwise, it sends back a NACK. This will continue until the original data packet is successfully sent or a preset number of retries is exceeded. Each node also uses *Collision Detection Multiple Access* (CDMA) to determine when to transmit on the network signal lines. It is akin to the old-fashioned telephone party line, where a user first listens to determine if anyone is talking and then starts talking if the line is free. In the case where a collision does occur, each potential sender "backs off" a random amount of time (tens of

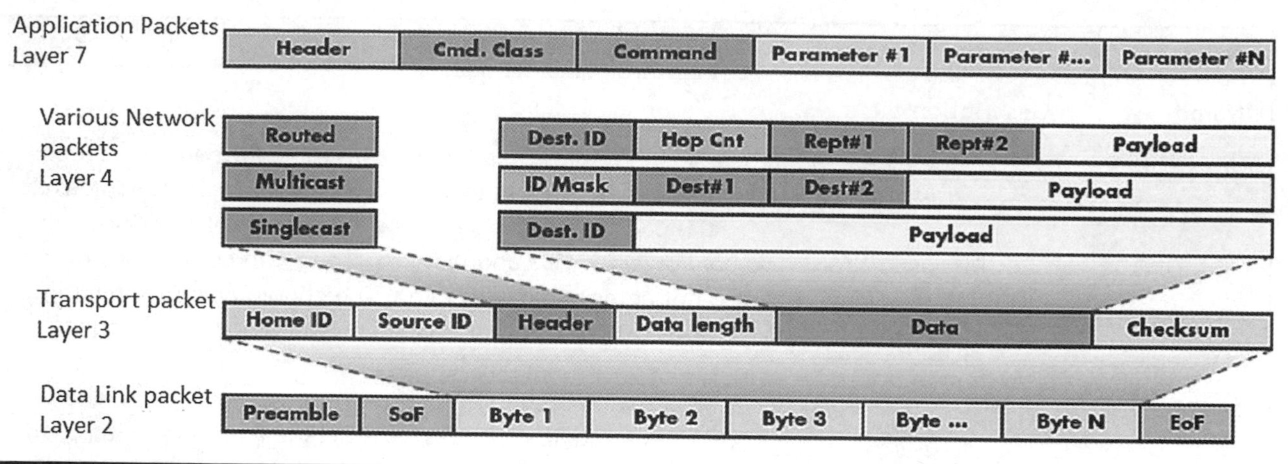

Figure 7-2 Z-Wave network packets versus model layers.

milliseconds) and then tries again. Using ACKs and NACKs depends upon an error-checking capability, which is why two checksum bytes are included in every data packet. These bytes are used by the receiving node in a mathematical algorithm to determine if the received data packet was corrupted during transmission. There are many good tutorials available on the Internet regarding checksums and error detection, if you are interested in digging further into the subject.

Layer 3 with the help of network Layer 4 further refines the packets by identifying what the packet type is, where it is going, and where it's been, and then sets the data payload. There are five packet types used in the Z-Wave network, which are described in Table 7-3.

Layer 4 sets up the routing, thereby ensuring that the packets are sent along the correct paths to reach the desired nodes. It also ensures that all nodes are repeating as configured. The routing configuration is maintained in a table stored in the primary network controller. It is possible to have multiple controllers on the network, but only one is designated as primary.

A simple network topology and descriptive routing table is shown in Fig. 7-3. A "1" in the routing table indicates a node-to-node logical connection.

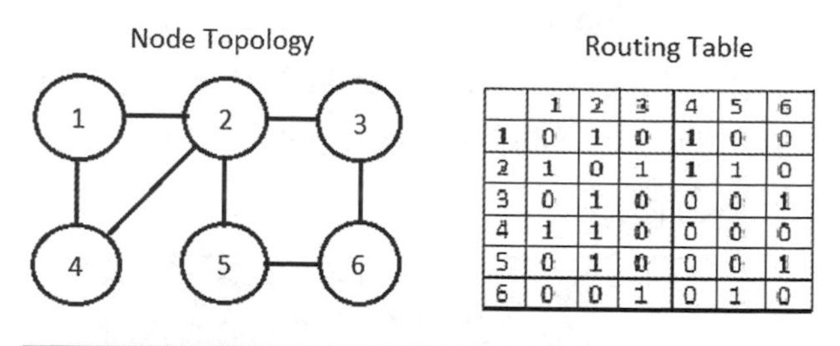

Figure 7-3 Network topology and routing table.

Table 7-3 Z-Wave Packet Types

Packet Type	Description	Payload Length	Remarks
Multicast	Broadcast to all network nodes	64 bytes max	0xff address
Singlecast	Sent to a specific node	64 bytes max	Up to 232 nodes
Routed	Repeated packet	64 bytes max	4 hops max
ACK	Acknowledgement	0	
NACK	Negative Acknowledgement	0	

Every Z-Wave network has a four-byte ID called the *Home ID*. Each primary controller has this Home ID that slave nodes acquire when they are joined to the network. Every secondary controller also uses this same ID when they are attached to the network. Individual slave nodes have a one-byte ID that is assigned by the primary controller when that node joins the network.

Network Devices

There are two main device types that make up a Z-Wave network: controllers and slaves. Slaves are also known as end-point devices because they can only respond to messages sent by the controllers. They typically have built-in microprocessors with GPIO pins that control components, such as TRIACs, which turn AC power on and off. Figure 7-4 shows a Z-Wave enabled duplex outlet. It would be installed in the same way as any ordinary duplex outlet would be installed, *except* that this outlet is also controllable using Z-Wave technology. The other big difference is that this outlet is much more expensive than a regular duplex outlet; however, you don't need many of them for most home automation setups.

Figure 7-4 Z-Wave enabled duplex outlet.

Notice the white button to the left center of the outlet in Fig. 7-4. This is what a user presses to join the device to the network, when prompted by the controller menu. This operation will be shown in a later section.

Controllers have two subtypes: *portable* and *static*. Portable controllers are in the form of remote controls, one of which is shown in Fig. 7-5. These controllers must be able to self-discover their location within the network topology because they would not be typically placed in a fixed location. Such self-discovery is made possible by having the portable controller "ping" nearby nodes that are within RF range. The controller can, thus, join the network based upon the ping results. Portable controllers are battery-powered for portability reasons and are usually used as primary controllers in Z-Wave networks.

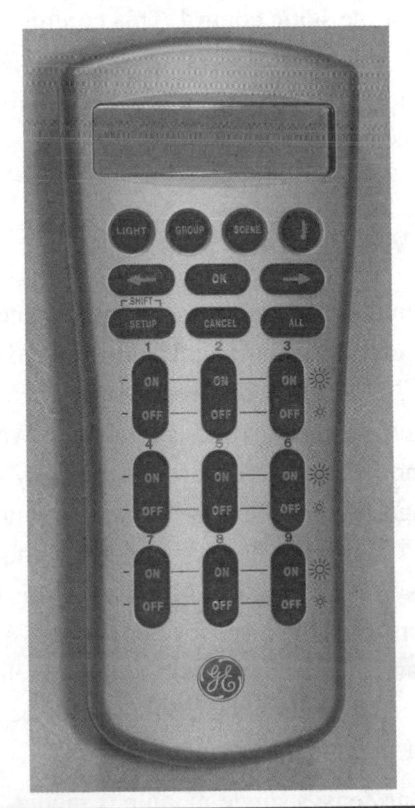

Figure 7-5 Portable Z-Wave controller.

The static controller is another subtype, and it so named because its logical location is fixed at the time of the initial network configuration. It is usually powered from the AC mains and constantly listens for network traffic. A static controller can serve as a secondary controller in an advanced network configuration. The current network configuration may be stored in it, and if so, it is known as a *Static Update Controller* (SUC).

More often, the role a static controller plays is to serve as a bridge between non Z-Wave components such as X-10 devices. In that configuration, the static controller serves as a virtual node between the X-10 device and the Z-Wave network. The network can have up to 125 virtual nodes that help incorporate older technologies into the modern Z-Wave system. Static controllers may also serve as TCP/IP gateways, thus allowing the network to connect to the Internet, if so desired. Finally, static controllers can serve as primary controllers in the configuration where the "normal" primaries act as proxies for the static control. This configuration type is known as an *SUC ID Server* (SIS). All told, there is an incredible amount of flexibility in Z-Wave network configurations.

The Z-Wave Chip

The original Z-Wave chip was designed and manufactured by Zensys, now known as Sigma Designs. All certified Z-Wave component manufacturers must use this authentic Z-Wave chip in their devices. This ensures that any Z-Wave node properly joins the network and communicates with other nodes produced by other manufacturers. The Zensys chip design is discussed in this section because it forms the basis for the whole Z-Wave concept and is important background to help you comprehend how the RasPi can function as a controller in the network.

A recent Zensys single module is model ZW3102N, containing a ZW0301 chip that uses the venerable 8051 core with a 32-MHz external crystal. This is a hybrid module containing a lot of additional components, including an RF transceiver operating on either the U.S. or European ISM frequency. There is also a built-in digital modem along with a hardware implementation of the network stack operations that were discussed in the previous section. The ZW0301 chip has only 32 kB of Flash memory and a meager 2 kB of SRAM. It operates on a supply voltage range of 2.1 to 3.6 V and consumes a maximum of 36 mA when transmitting. Figure 7-6 is the block diagram of the ZW3102N, showing all the components that constitute this module.

The ZW0301 microcontroller chip discussed above is also shown in the block diagram. It has several of the standard functions that have been discussed in previous chapters, including the SPI and UART serial interfaces. The chip also has timers, interrupts, a watch-dog monitor, power management, and brownout detection. It has a four channel, 12-bit ADC, a pulse-width-modulation controller, and an enhanced TRIAC control that has zero crossing detection. A total of 10 GPIO lines are available, but some are multiplexed or shared with other I/O functions.

The ZW3102N module is very small; Fig. 7-7 is a photo of it compared to a U.S. 25-cent coin.

The module does need an external antenna and a few capacitors and inductors to complete a Z-Wave device installation. The software is fixed in the flash memory and is not available for examination or modification. This is where this RasPi project will open up the Z-Wave network so that you have a chance to experiment with various configurations and monitor network traffic. But first, I would like to demonstrate a simple Z-Wave network that uses a portable controller (Fig. 7-5) along with two nodes, one being the duplex outlet (Fig. 7-4) and the other being an outdoor module shown in Fig. 7-8. Notice the black button located on the top of the device in Fig. 7-8. The user presses it to join

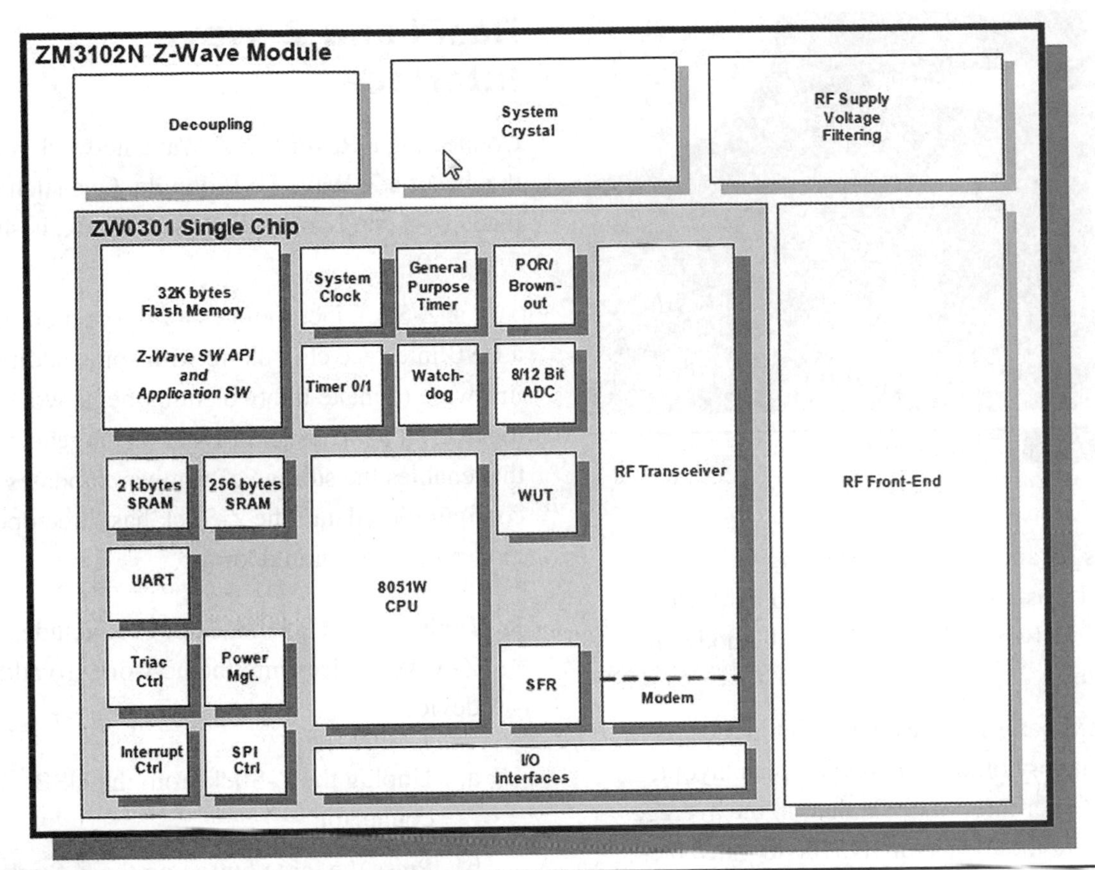

Figure 7-6 Block diagram of the ZW3102N Z-Wave module.

the device to the network when prompted by the controller menu.

Notice the black button located on the top of the device in Fig. 7-8. The user presses it to join the device to the network when prompted by the controller menu.

My test network slaves will be made up of the duplex outlet and the outdoor module, each controlling a small table lamp. The duplex outlet will actually be connected to a power cord plugged into a regular outlet for this temporary test arrangement. Figure 7-9 shows the test setup on my dining room table.

At first, I arbitrarily assigned a device number 4 to the duplex outlet and a device number 8 to the outdoor module. I then proceeded to turn

Figure 7-7 ZW3102N module compared to a U.S. 25-cent coin.

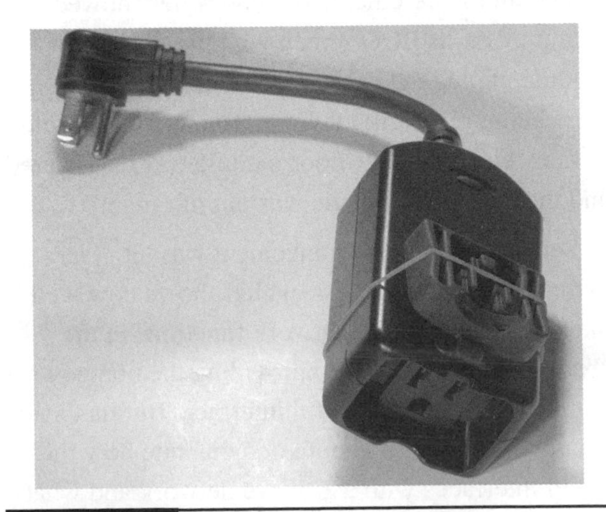

Figure 7-8 Z-Wave outdoor module.

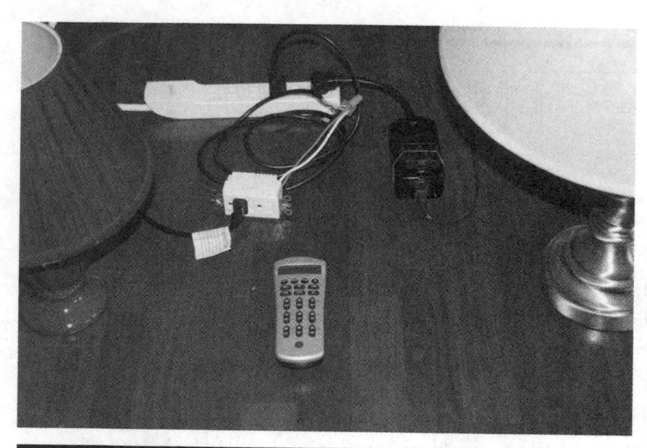

Figure 7-9 Z-Wave test system.

the lamps on and off, and everything worked just fine. I was also able to control both devices simultaneously by selecting the "All" mode on the remote.

The next part of the test was a bit harder, as I have a smaller home with an open plan layout, meaning fewer interior walls than in the average cape-style home. I was finally able to place the outdoor module device in the basement and the duplex outlet on the first floor and then operated the controller in a second floor bedroom. I was not able to turn on the basement module without having the first-floor module plugged in. This proved that the first-floor module was digipeating and forwarding the control packets to the module located in the basement. The controller showed "Failure" on its LCD screen with the first-floor module unplugged, which indicated that no ACKs were being received. Obviously, no NACKs could be sent, since the first-floor module was unpowered and the basement module was out of range.

Setting up the test arrangement was an extremely simple process, which shows how well the high-tech Z-Wave network functions in the "background," while also providing the homeowner with a very easy and useful interface. But that's not what we are after. I want to demonstrate how the RasPi interfaces with a Z-Wave network and what "neat" experiments can be accomplished.

RasPi and Z-Wave Interface

Connecting a RasPi to a Z-Wave network requires the use of a Z-Wave USB dongle. One such device made by Aeon Labs, called the Z-Stick, is shown in Fig. 7-10.

The Z-Stick incorporates a Zensys module and a USB interface chip along with some additional firmware to make the two components work together. It also has an internal rechargeable battery that enables the storage of firmware updates and configuration data. The Z-Stick has three operating modes that you should know:

1. **Inclusion**—This mode adds or includes Z-wave devices into the network. To add a device:

    a. Unplug the Z-Stick from the USB connector.

    b. Press the large button on the Z-Stick. The Z-Stick LED will start to blink slowly.

    c. Go to the device that you wish to add (while continuing to press the large Z-Stick button) and press and release the device's button.

    d. The Z-Stick LED will blink rapidly for several seconds, then glow steadily for three seconds, and finally return to a slow blinking state. The device has been added to the network.

Figure 7-10 Aeon Labs Z-Stick.

2. **Removal**—This mode will remove or exclude Z-wave devices from the network. To remove a device:

   a. Unplug the Z-Stick from the USB connector.
   b. Press and hold the large button on the Z-Stick for about three seconds. The Z-Stick LED will start to blink slowly and then transition to a fast blink.
   c. Go to the device that you wish to remove (while continuing to press the large Z-Stick button) and press and release the device's button.
   d. The Z-Stick LED will then glow steadily for three seconds and finally return to a fast blinking state. The device has been removed from the network.

3. **SerialAPI**—This is the mode where the Z-Stick acts as the portal between the RasPi and the Z-Wave network. Simply plug it into a powered-hub USB connector. The RasPi probably does not have sufficient power for the Z-Stick. The RasPi software will now take control of the Z-Wave network.

I now have to take a brief detour from the Z-Wave to introduce the SSH login process, since I use that in establishing the control software environment.

SSH Login

In this section, I will show you how to log into the RasPi by using a network connection, as I mentioned in Chap. 5. The Wheezy Linux distribution, as well as many others, includes a great service known as SSH, short for *Secure Shell*. It is a network protocol that uses cryptographic means to establish secure data communication between two networked computers connected via a logical, secure channel over a physical, insecure network. SSH uses both server and client programs to accomplish the connection.

One of the questions you will be asked when first configuring your RasPi is whether or not to start sshd upon bootup. I recommend that you answer "yes" as that automatically starts the SSH daemon each time you start the RasPi. The second portion of the connection is the client program, which is highly dependent upon what type of computer you are using to connect to the RasPi. I recommend using putty.exe, since most readers will be using a Windows®-based machine. putty is freely available from a variety of Internet sources, so I would recommend a Google search to locate a good download mirror.

You should see the Fig. 7-11 screenshot, assuming that you answered "yes" to the sshd question and have downloaded and are running putty on a Windows-based computer connected to the same network that connects to the RasPi. Don't be concerned with the host name that appears in the screenshot; I will get to that shortly.

When you click on the Open button at the bottom of the putty screen, you will see in Fig. 7-12, a screenshot of a RasPi terminal window asking, in this case, for a login password.

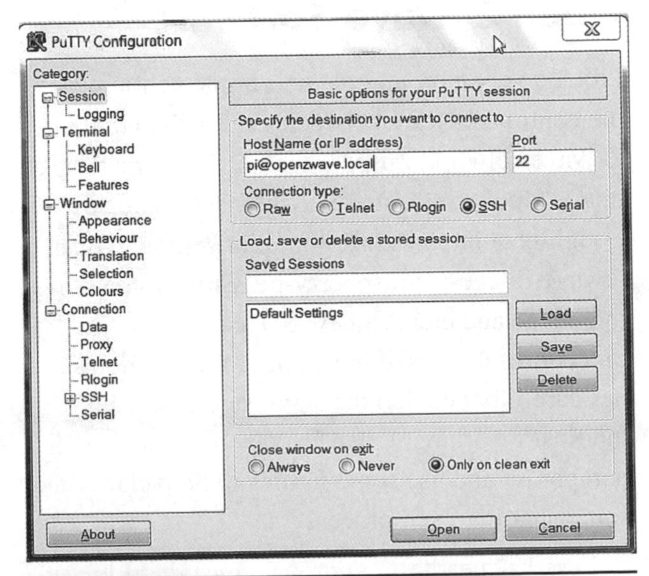

Figure 7-11 putty screenshot.

Figure 7-12 Raspberry Pi terminal window.

At this point, you are in a RasPi terminal window session. It is absolutely no different from looking at a monitor connected directly to the RasPi and using a locally connected keyboard and mouse. This transparency is what makes SSH so great—it allows you to login remotely to the RasPi without being concerned with any minutia about the connection. You may type in any normal command and have the RasPi respond as appropriate.

I will now return to the Z-Wave control software discussion, now that you are a bit familiar with SSH.

Open Z-Wave Software

I took a very easy approach to implementing the control software between the RasPi and the Z-Wave network. I used a prebuilt Wheezy Raspian image kindly made freely available by Thomas Loughlin at http://thomasloughlin.com/new-open-zwave-image-for-raspberry-pi/. Simply download the image and create a new SD, card using the procedures discussed in Chap. 1. While Thomas has carefully detailed the steps required to get the RasPi working with a Z-Wave network, I will paraphrase and use some figures to help clarify the somewhat involved process.

I used a "headless" approach, meaning I logged into the RasPi using putty. Headless is a bit of network jargon meaning you don't need a local monitor or keyboard for the RasPi because you will login remotely using the RasPi as a server. The step-by-step procedure for starting the Z-Wave RasPi network is:

1. First add one or two devices to the network as described above. This establishes a beginning configuration that enables you to see and control some devices.
2. Plug the Z-Stick into a powered hub connected to an unpowered RasPi that has the downloaded Open Z-Wave image.
3. Power on the RasPi and wait a minute or so. Remember, we are running headless and are not looking at a monitor screen connected to the RasPi.
4. In the opening putty screen, enter

   ```
   pi@openzwave.local
   ```

   as a host name. (Fig. 7-11)
5. The password is the regular one,

   ```
   raspberry
   ```

6. A terminal window (Fig. 7-12) will appear.
7. You are now ready to take one of two approaches to monitor and control the Z-Wave network.

lightscontrol Server

One of two server applications may be run at this point. The first is a web app named "lightscontrol" that was created by Conrad Vassallo and can be downloaded from the website, http://conradvassallo.com/category/open-zwave-controller/. This will be the one that I use. The second approach is a bare bones server app that Thomas created to test the Z-Wave network. I will discuss that server after demonstrating the lightscontrol server. All you need to do to start this service is type the following at the command line prompt:

```
lightscontrol ↵
```

This starts the lightscontrol server. Figure 7-13 shows the lengthy preamble to the lightscontrol server web app.

Now, you should minimize the terminal control window, keeping in mind that there is still an underlying active SSH connection that will allow you to input commands to the RasPi. We are going to need putty after experimenting with the first web server. Next, open a browser and enter this for the URL:

```
pi@openzwave: ~
Using username "pi".
pi@openzwave.local's password:
Linux openzwave 3.2.27+ #250 PREEMPT Thu Oct 18 19:03:02 BST 2012 armv6l

The programs included with the Debian GNU/Linux system are free software;
the exact distribution terms for each program are described in the
individual files in /usr/share/doc/*/copyright.

Debian GNU/Linux comes with ABSOLUTELY NO WARRANTY, to the extent
permitted by applicable law.

##########################################################
Hostname: openzwave.local

Starting open-Zwave Servers:

To start the default server type: lightscontrol
Then go to http://openzwave.local/ in your browser.
U: Admin P:admin

You must have the "lightscontrol-server" running before accessing the
web interface.  On the first use, go to devices, then "Import".

To kill the "lightscontrol-server", type killOZW.

To start the "basic-server", type basic and go to
http://openzwave.local/basic/ to control your lights.

Both open-Zwave servers cannot be running at the same time.  If a server will
not stay running, reboot.

The command "killOZW" will also kill the "basic-server".  I normally wait a
few seconds between each start / stop and a few seconds before loading the
web interface.

The command "webroot" will take you to the nginx root directory.

To edit this message (or make it go away): sudo nano /etc/motd
##########################################################
Last login: Wed Dec 12 23:31:37 2012
pi@openzwave ~ $ lightscontrol
[1] 2236
pi@openzwave ~ $ nohup: ignoring input and redirecting stderr to stdout

pi@openzwave ~ $ 
```

Figure 7-13 lightscontrol web server preamble screenshot.

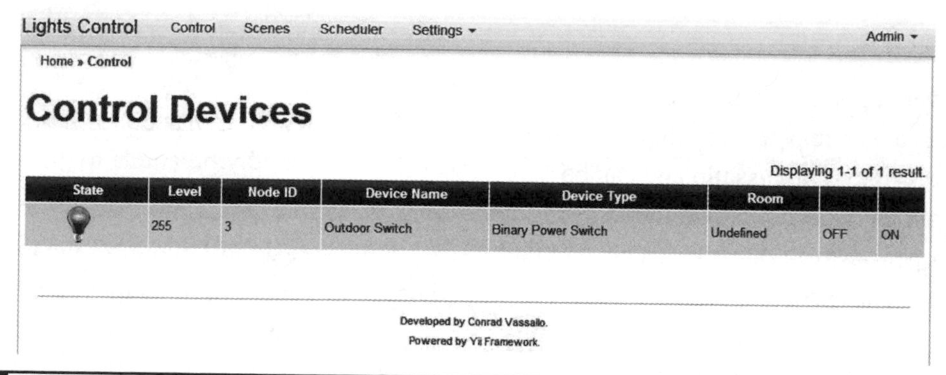

Figure 7-14 lightscontrol server page.

```
http://openzwave.local ↵
```

This should take you to the initial lightscontrol server page, as shown in Fig. 7-14.

In my case, I had initially set up one device, a GE outdoor power module controlling a lamp, as I discussed earlier in the chapter. Table 7-4 shows the web page attributes for this device.

Clicking on the web page ON and OFF buttons turned the corresponding Z-Wave controlled lamp on and off, thus confirming that all the software and network connections were properly operating.

I have included some screenshots of the web app configuration pages to help illustrate how to configure a Z-Wave network to support typical home automation tasks. There is no professional documentation provided with this open-source project, which is often the case, but I am not complaining and am completely appreciative of the tremendous unpaid effort that goes into such project development done for the general good.

Figure 7-15 is a screenshot of the Scenes web page where one or more devices can be grouped collectively to support a common event happening in the home, such as Wake Up, TV, Kids Sleeping, etc.

Figure 7-16 is a screenshot of the Scheduled Events page that works in conjunction with the Scenes page, where the actual device times that support a specific scene may be programmed.

My recommendation is to simply experiment with the web app, determining the appropriate inputs to the various pages in order to execute the

Table 7-4 Z-Wave Device Attributes

Attribute	Value	Description
State	n/a	A bulb icon that changes from red to gray, for ON or OFF respectively
Level	255	
Node ID	3	Network ID
Device Name	Outdoor Switch	Enumerated from device table
Device Type	Binary Power Switch	Enumerated from device table
Room	Undefined	Automatically assigned for undefined room
OFF	n/a	An action button that turns device off
ON	n/a	An action button that turns device on

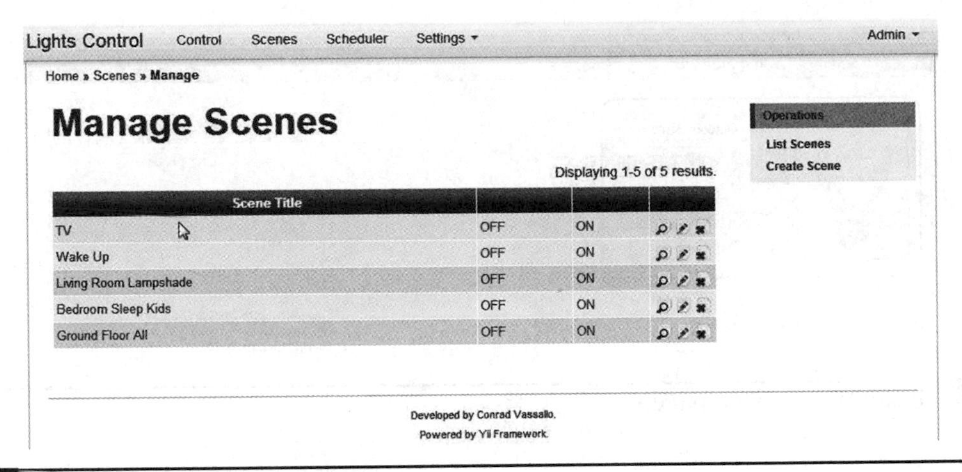

Figure 7-15 Scenes web page screenshot.

desired automation sequences. Figuring out the web page inputs should not be a very hard task for you given that you already have reached this stage in building a RasPi home automation system and have successfully loaded and run all the support software. Mr. Vassallo has stated in his blog that the software is still evolving, so don't be surprised if a better, perhaps somewhat easier, version is developed and put in the public domain.

"basic" Server

The second approach mentioned above is to use the "basic" server that Mr. Loughlin created to test the Z-Wave network functions. In order to do this, you must first kill the lightscontrol server that is currently running, since there cannot be two Z-Wave servers running concurrently. You can also close the browser at this time; it will need to be reopened, however, once the next service is started. Kill the current web service by typing

```
killOZW ↲
```

at the SSH command line. This is the reason why I did not want you to stop running putty.

To start the "basic" server, simply enter the following at the command line:

```
basic ↲
```

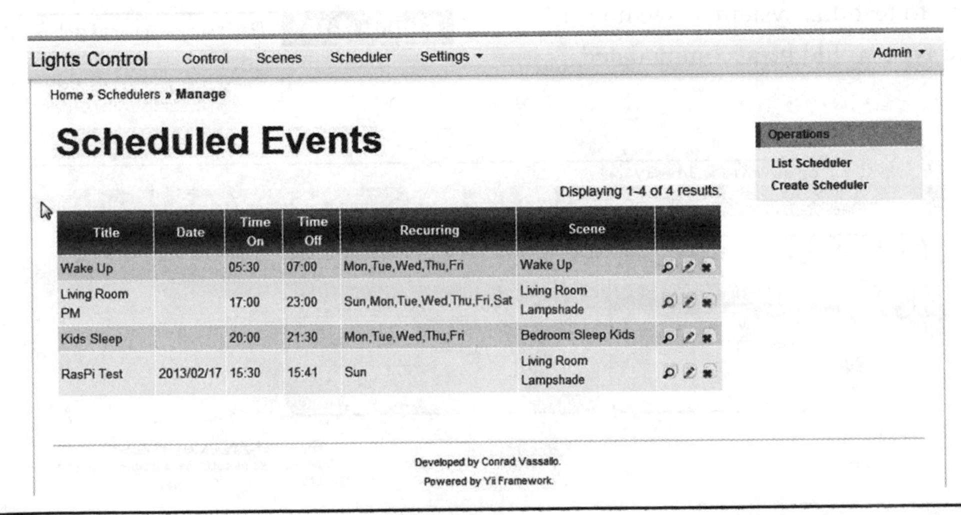

Figure 7-16 Scheduled Events web page screenshot.

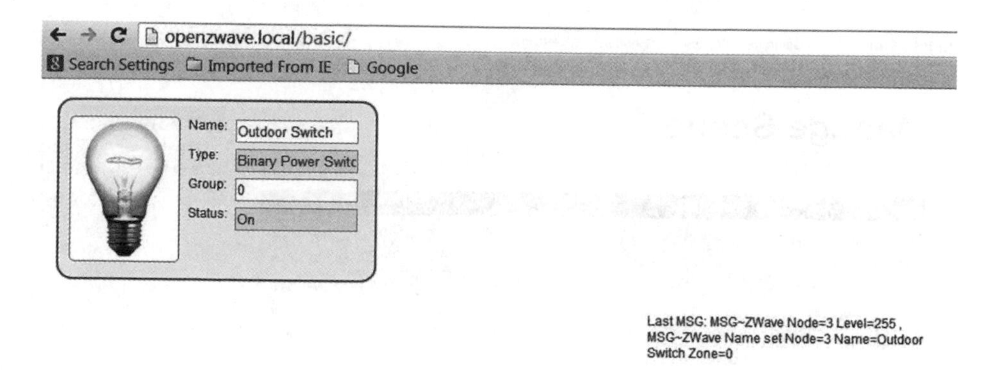

Figure 7-17 "basic" server web page.

Once that is done, minimize putty and open the browser going to this website:

```
http://openzwave.local/basic/
```

This should take you to the "basic" server page, as shown in Fig. 7-17.

All devices that have been added to the Z-Wave network and stored in the Z-Stick memory will be displayed on this page. As you can see, the only device displayed is the outdoor power module that was shown in the previous web server demo. The light may be turned on and off by clicking on the icon. There is no provision for setting Scenes or preprogrammed times, as this is just a very functional web test app. The essential device attributes are shown as text in the middle of the page.

I next wanted to test this system to see how a different device type would be accommodated.

To do this, I shut everything down and then added a new device to the network. This was a battery-operated, Z-Wave enabled motion sensor, as shown in Fig. 7-18.

I next restarted the "basic" web service and loaded the appropriate web page. Figure 7-19 shows this new page with both the outdoor module and the motion sensor displayed.

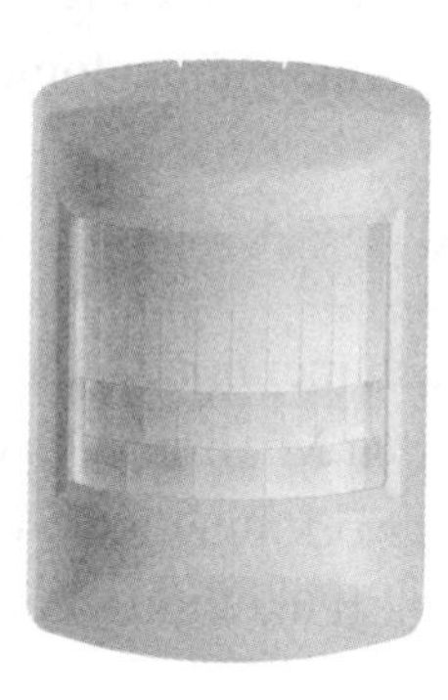

Figure 7-18 Battery-operated, Z-Wave motion sensor.

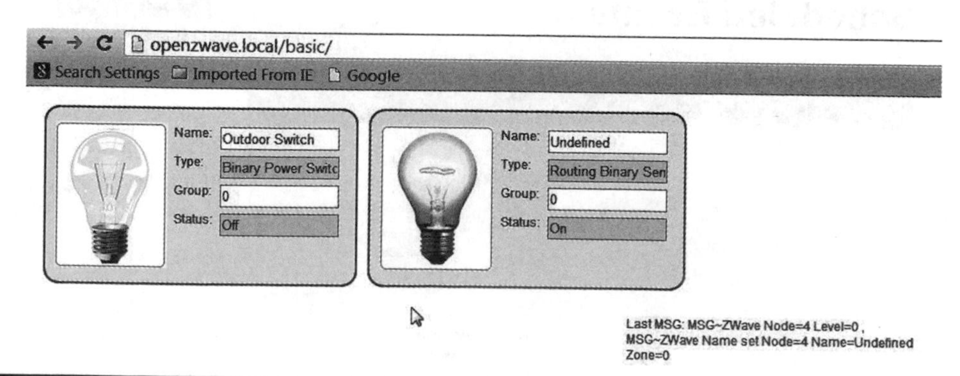

Figure 7-19 Revised "basic" server web page.

The motion sensor sends data back to the Z-Stick, which I confirmed by waving my hand in front of the sensor and observing that the state shown on the sensor window changed from off to on. Note, I did have to reload the page to observe this change. Also, the essential device attributes shown on the page reflect the selected device.

Summary

I started this chapter with a definition of home automation and provided a list of representative tasks that would fall under the broad umbrella that comprises home automation.

Next followed a discussion of the underlying base technologies that support home automation implementations. I provided a rationale for selecting the Z-Wave protocol: it is a modern, highly flexible system that is very easy to configure and lends itself quite well to a simple RasPi interface connection.

Two representative Z-Wave devices were next described along with a working, small-scale Z-Wave network. Then we conducted a network demonstration by using a commercial remote control device.

The core Z-Wave chip was discussed to provide you with a better understanding of how the overall network functions and how devices are highly dependent upon the chip functions, which can be added seamlessly to the network.

The Z-Stick that enables the actual RasPi to Z-Wave interface was shown along with the driver software. However, before discussing the software, I showed you the highly useful SSH remote login procedure.

The open Z-Wave software package was shown with two variations for the web application side: 1) lightscontrol and 2) "basic."

Some simple tasks such as turning a lamp on and off were demonstrated as well as the ability to sense motion in a room.

CHAPTER 8

Home Security Controller

Introduction

For this chapter's project, you'll learn how to connect a webcam to the RasPi so that a homeowner can view a selected home area when an intrusion signal is detected by the RasPi. A laser trip device serves as the intrusion sensor. When the RasPi detects an intrusion signal, the RasPi will send an e-mail notification to a preset e-mail address to notify the homeowner that the laser has tripped, thus indicating that a possible intruder is on the premises. The homeowner will then open a browser, go to a website where the webcam video is available, and examine the ongoing activity. The homeowner could then determine whether to call the police or simply to disregard the alarm, if it had been innocuously tripped.

Motion is the name of the software package installed on the RasPi. It is a very sophisticated suite of vision applications that provides the project's webcam monitoring function. Many other features are also available and will be discussed in the software section.

Webcam Selection

The webcam used in this project is a fairly new Logitech C920, as shown in Fig. 8-1.

The webcam has a high definition camera capable of producing excellent videos; however, the software used with it in this project will limit its performance. With that in mind, note that many different webcams may be used. In addition, the video requirements are quite low, and the Linux Raspian distribution will automatically detect quite a few types of webcams, including many older ones. If you already own a webcam, my suggestion is to plug it into a USB port and then type the following into a command line:

```
lsusb ↵
```

Figure 8-2 is a screenshot of the command's output. Device 008 is the webcam and the other Logitech device listed is a keyboard. If in doubt about which device is the right one, when multiple devices from the same manufacturer show up, simply unplug the device and rerun the command to see which one disappears.

One key point I discovered in creating this project is that the webcam has to be directly plugged into one of the two RasPi USB ports. The webcam was not detected by the software when it

Figure 8-1 Logitech C920 webcam.

```
pi@raspberrypi: ~
File  Edit  Tabs  Help
pi@raspberrypi ~ $ lsusb
Bus 001 Device 002: ID 0424:9512 Standard Microsystems Corp.
Bus 001 Device 001: ID 1d6b:0002 Linux Foundation 2.0 root hub
Bus 001 Device 003: ID 0424:ec00 Standard Microsystems Corp.
Bus 001 Device 004: ID 046d:082d Logitech, Inc.
Bus 001 Device 005: ID 05e3:0608 Genesys Logic, Inc. USB-2.0 4-Port HUB
Bus 001 Device 006: ID 7392:7811 Edimax Technology Co., Ltd EW-7811Un 802.11n Wi
reless Adapter [Realtek RTL8188CUS]
Bus 001 Device 007: ID 1241:1177 Belkin F8E842-DL Mouse
Bus 001 Device 008: ID 046d:c326 Logitech, Inc.
pi@raspberrypi ~ $ 
```

Figure 8-2 lsusb output screenshot.

was connected to a powered-hub USB connector. I am not sure why this happened but it may be related to an issue with the hub power supply.

Protected Security Zone

Figure 8-3 is an actual image capture from the webcam. It shows the protected zone, which, in this case, is the exit to my home's back deck. I will discuss how images may be captured later in the software section. It is now important to show the intended protection zone because it greatly affects how the laser trip assembly will be deployed as part of the overall security scheme.

Figure 8-3 Protected zone.

Laser Trip Assembly

I reused the same laser trip device that I used in the camera controller project. Figure 8-4 shows this assembly mounted on a single piece of Lexan. The only difference in this trip setup is that I used a mirror to reflect the beam back to the detector's phototransistor when an object interrupts the beam between the laser pointer and the detector in the protected zone. The height at which you mount the laser and the detector board is your choice,

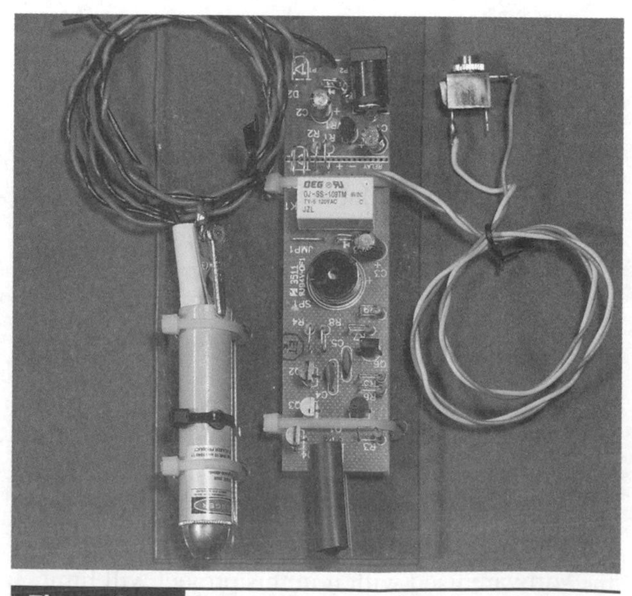

Figure 8-4 Laser trip assembly.

Figure 8-5 Operational laser trip assembly.

but I found that about four inches above the floor worked quite nicely. The total path length of the laser beam was about 14 feet, which was not an issue with this system.

Figure 8-5 shows the laser trip assembly operating in the evening. This photo was taken with a flash to clearly show all the components, including the mirror to the right.

I took another photo, but this time in almost total darkness. In Fig. 8-6, you can see the emitted laser beam and a dot of reflected light in the mirror; however, you cannot see the projected laser beam. This, of course, conflicts with what you see in the movies when the hero (or bad guy) attempts to get past the laser beams protecting the secret vault or whatever. I would actually have to put smoke or dust in the beam's path to see it, which would have caused much consternation in my household.

The extra red light seen on the detector side is the returned beam reflecting off of the nearby white trim board. You could place some black cloth behind the detector if a more stealthy installation is desired.

The trip system functioned very well, never missing my exiting or returning. Another trip assembly could be stationed at four feet above the

Figure 8-6 An elapsed-time photo of the laser trip assembly in operation.

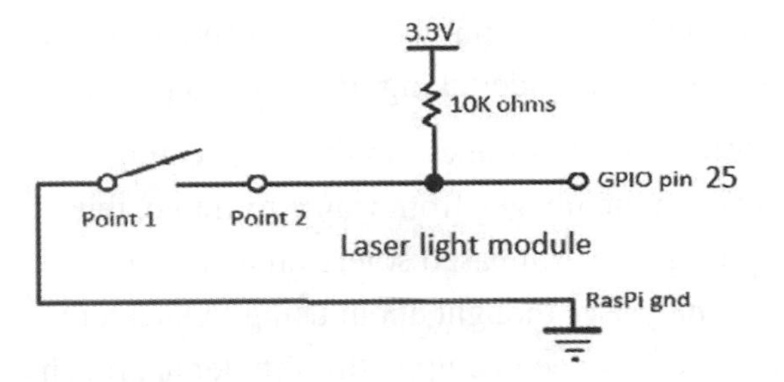

Figure 8-7 Schematic of relay switched interface for the laser trip assembly.

floor if you are worried about intruders simply stepping over the invisible beam. My philosophy regarding such matters is that criminals will simply go to easier targets if they suspect or detect that a relatively sophisticated intrusion system is in place *and* operational.

RasPi Interface to the Laser Trip Assembly

I have used the same relay switched circuit that was used in the camera control project. The schematic is shown in Fig. 8-7.

The circuit is extremely simple, yet highly reliable—two attributes that are important in security systems. The relay contacts are normally open, which means that the GPIO pin will be in a high state due to the 10 kΩ resistor tied to 3.3 V. The pin will go to ground or to 0 V when the detector is tripped. The software, thus, has to check periodically for a low state on the selected GPIO pin and then initiate all the desired actions, which will be discussed in the software section.

Motion Software Package

I selected a software package named Motion to enable remote viewing of the webcam. This is a comprehensive package that contains many features, far more than could be covered in this chapter. Being able to use this package from among the others available is why the RasPi is so useful. Creating similar software for a more traditional board, such as one from the Arduino series, would be a substantial undertaking, if it is even possible.

Motion also has the capability of detecting changes in the images from frame to frame, thus triggering an alarm based solely on real-time video analysis. I thought about using this feature but eventually decided upon the simpler approach of using an external trip system. Detecting an intrusion event based on the video alone requires the RasPi to be almost solely committed to vision processing, which could affect the availability of the video for remote viewing. This may have been an unwarranted decision, but remember that the Motion software was created years before the RasPi came on the scene and probably was based on using much higher-clock-speed PCs.

The key feature that is used from the Motion package is the built-in web server. This server receives the video stream from the webcam and sends it off in TCP/IP format over a predefined port. All you need to remotely view the webcam video is a browser pointed to the RasPi IP address and port number, nothing more. This feature makes the viewing exercise extremely simple. But there is more: Motion provides for using more than one webcam. You can set up multiple webcams, each with its own port number, so that it can monitor multiple locations throughout the home. Each webcam video feed is handled by what is known as a *thread* within the Motion software. I saw provisions for four threads in the Motion configuration file from which I presume four webcams could be handled. However, I seriously wonder if the RasPi has the processing power to manage four simultaneous video feeds. In any case, this project is only concerned with one feed that I know works very well.

Motion Features

Motion has a substantial number of features that enable it to accomplish an amazing number of functions. The user manual is on-line at http://www.lavrsen.dk/foswiki/bin/view/Motion/WebHome, and it is over 100 pages in length. It would take a complete book to document all the features. However, the configuration file itself contains many self-documenting comments that should help you explore some of the additional features of this software package.

A portion of the configuration file dealing with motion detection is listed on page 107 to illustrate what I mean.

Motion Setup

You must, of course, install the Motion package before using it. I would strongly suggest that you update and upgrade your distribution before installing Motion. Simply type the following at a command line prompt to update and upgrade the Linux distribution in use:

```
sudo apt-get update ↵
sudo apt-get upgrade ↵
```

Be patient, since the updates and upgrades can take a bit of time if there are many to install.

Next install Motion by typing:

```
sudo apt-get install motion ↵
```

Again, be a bit patient as this package is over 50 MB in size and has many component parts.

Motion will be run in the background as a daemon, which means that it will be a constantly available service. To enable the daemon, you must edit the /etc/default/motion file. Type the following:

```
sudo nano /etc/default/motion
```

You will see in the nano editor the line:

```
start_motion_daemon=no
```

Motion Detection Settings

```
# Threshold for number of changed pixels in an image that
# triggers motion detection (default: 1500)
# threshold 1500
# CHANGED
threshold 4000
# Automatically tune the threshold down if possible (default: off)
threshold_tune off
# Noise threshold for the motion detection (default: 32)
noise_level 32
# Automatically tune the noise threshold (default: on)
noise_tune on
# Despeckle motion image using (e)rode or (d)ilate or (l)abel (Default: not defined)
# Recommended value is EedDl. Any combination (and number of) of E, e, d, and D is
valid.
# (l)abeling must only be used once and the 'l' must be the last letter.
# Comment out to disable
despeckle EedDl
# Detect motion in predefined areas (1 - 9). Areas are numbered like that: 1 2 3
# A script (on_area_detected) is started immediately when motion is     4 5 6
# detected in one of the given areas, but only once during an event.    7 8 9
# One or more areas can be specified with this option. (Default: not defined)
; area_detect value
# PGM file to use as a sensitivity mask.
# Full path name to. (Default: not defined)
; mask_file value
# Dynamically create a mask file during operation (default: 0)
# Adjust speed of mask changes from 0 (off) to 10 (fast)
smart_mask_speed 0
# Ignore sudden massive light intensity changes given as a percentage of the picture
#area that changed intensity. Valid range: 0 - 100 , default: 0 = disabled
# CHANGED
lightswitch 0
# Picture frames must contain motion at least the specified number of frames
# in a row before they are detected as true motion. At the default of 1, all
# motion is detected. Valid range: 1 to thousands, recommended 1-5
# CHANGED
minimum_motion_frames 5
# Specifies the number of pre-captured (buffered) pictures from before motion
# was detected that will be output at motion detection.
# Recommended range: 0 to 5 (default: 0)
# Do not use large values! Large values will cause Motion to skip video frames and
# cause unsmooth mpegs. To smooth mpegs use larger values of post_capture instead.
# CHANGED
pre_capture 10
# Number of frames to capture after motion is no longer detected (default: 0)
# CHANGED
post_capture 100
# Gap is the seconds of no motion detection that triggers the end of an event
```

```
# An event is defined as a series of motion images taken within a short timeframe.
# Recommended value is 60 seconds (Default). The value 0 is allowed and disables
# events causing all Motion to be written to one single mpeg file and no pre_capture.
# CHANGED
gap 5
# Maximum length in seconds of an mpeg movie
# When value is exceeded a new mpeg file is created. (Default: 0 = infinite)
max_mpeg_time 0
# Always save images even if there was no motion (default: off)
output_all off
```

Change the "no" to "yes," then save the nano buffer and exit the editor.

Next comes Motion's configuration file. Motion has no GUI; it is totally configured by making changes to its configuration file, /etc/motion/motion.conf. This is a very big text file—well over 600 lines—although much of file's contents are comments inserted to help the user. Fortunately only a few changes are necessary for this project. In Table 8-1, I provide the changes to be made to the configuration file section, but I do not give you step-by-step instructions because it should be fairly obvious by now how the editor functions.

Again, start a nano editor session as follows:

```
sudo nano /etc/motion/motion.conf ↵
```

Save the changes and exit the nano editor. Now, you must start the Motion server, which is done by typing:

```
sudo service motion start ↵
```

One nice feature of having the Motion web server running as a daemon is that it is automatically started each time you boot the RasPi. You may also stop or restart the service by typing:

```
sudo service motion stop ↵
sudo service motion restart ↵
```

That's it for the changes to make in the configuration file. I do want to briefly discuss why these changes were made. The change from daemon off to daemon on is obvious—it was needed to run Motion as a daemon. The next change—to make the port number 8081—is a bit historical, since the Motion web service has traditionally been assigned to this port. It is not a required port number and you can easily change it to any number that you desire as long as it is greater than 1024 and less than 65535. This range avoids the "well known ports" and goes to the maximum possible port number. My recommendation is to leave it at 8081.

The next change concerns localhost operation. Localhost refers to the same machine that is

Table 8-1 Motion Configuration File Changes

Section	Change From	Change To
Daemon	daemon off	daemon on
Live Webcam Server	webcam_port 0	webcam_port 8081
Live Webcam Server	webcam_localhost on	webcam_localhost off
HTTP Based Control	control_port 0	control_port 8080
HTTP Based Control	control_localhost on	control_localhost off

hosting the Motion web server. No other system can access the webcam if you restrict the service to localhost; hence it must be turned off.

The next two changes are similar to what was discussed above but are concerned with the remote control functions of the webcam. I do not enable any remote control functions, such as webcam panning or tilting in this project, but it is certainly doable. I suggest, however, that you simply keep the port number as assigned and disable the localhost operation.

Webcam Viewing

It is now time to test the remote webcam viewing functionality. You will need a separate computer on the same network that the RasPi is connected to. You will also need the RasPi IP address. It does not matter if the RasPi is connected via an Ethernet cable or by a Wi-Fi wireless adapter. My suggestion is to login into your network router and click on "Attached Devices." The local IP address for the RasPi should appear in the appropriate list. My Netgear router has separate lists for wired and wireless devices connected to the network. Yours might be slightly different but should show something similar to Fig. 8-8.

The entry RASPBERRYPI has the IP 192.168.1.21 on the wired portion of my home network. This is all I need to remotely view the webcam. I just have to type 192.168.1.21:8081 into a browser on another networked computer to view the real-time webcam video stream. Figure 8-9 is a screen capture of the video feed from the RasPi webcam. I used the robot car from Chaps. 12 and 13 as the subject.

Figure 8-9 Video stream screen capture.

E-Mail Notification

The next step in this project is to discuss the software that sends out an e-mail when the laser trip system is triggered. You will need a working Gmail account in order to have this software function because it uses a Python package named *smtplib*, which, in turn, uses Google's freely available Gmail smtp server. The software is also designed to run in a home network using a router. I will show you how to obtain the public facing IP address that is dynamically assigned by your Internet Service Provider (ISP).

The Python code that sends out the e-mail notification is shown on page 110.

Wired Devices

#	IP Address	MAC Address	Device Name
1	192.168.1.13	00:1D:BA:30:1A:42	<unknown>
2	192.168.1.28	00:26:2D:03:61:9C	
3	192.168.1.5	00:10:75:06:0D:6A	BA-060D6A
4	192.168.1.19	00:10:75:1A:0D:50	<unknown>
5	192.168.1.21	B8:27:EB:BC:74:67	RASPBERRYPI
6	192.168.1.3	00:0D:4B:71:73:8C	<unknown>
7	192.168.1.22	00:21:5A:0C:A0:E7	HP0CA0E7
8	192.168.1.20	00:90:A9:6F:8E:CB	MYBOOKWORLD

Figure 8-8 Router attached devices list.

Email Notification Python Code

```
import subprocess
import smtplib
import socket
from email.mime.text import MIMEText
import datetime
import os
#put the destination e-mail address here
to = 'youremail@isp.com'
#put your user gmail account here
gmail_user = 'user@gmail.com'
#put your gmail password here
gmail_password = 'password'
#everything is verbatim from this line on down
smtpserver = smtplib.SMTP('smtp.gmail.com', 587)
smtpserver.ehlo()
smtpserver.starttls()
smtpserver.ehlo()
smtpserver.login(gmail_user, gmail_password)
today = datetime.date.today()
arg = 'ip route list'
proc = subprocess.Popen(arg, shell=True, stdout=subprocess.PIPE)
data = proc.communicate()
split_data = data[0].split()
local_ip = split_data[split_data.index('src') + 1]
public_ip = os.system('wget http://ipecho.net/plain -O - -q > test.txt; echo')
public_ip = open('test.txt', 'r').read()
msg_content = 'Alarm detected. Public ip is %s:8081, local ip is %s' % (public_ip,
local_ip)
msg = MIMEText(msg_content)
msg['Subject'] = 'Raspberry Pi Alarm on %s' % today
msg['From'] = gmail_user
msg['To'] = to
smtpserver.sendmail(gmail_user, [to], msg.as_string())
smtpserver.quit()
```

Figure 8-10 shows a slightly altered screenshot of the e-mail received after I ran this script (the names and public IP are changed to protect my privacy).

I wish to mention two items regarding this program. First, the public IP is determined by using an external website, http://ipecho.net. There is no guarantee that this website will be operational for the indefinite future. If it becomes unavailable, then alternate sites should be used, and appropriate changes made to this program to accommodate the new site.

The second item is that I have not yet included the code for the GPIO trigger. This inclusion will consist of restructuring the program as a module that will be called when the laser trip activates the selected GPIO pin.

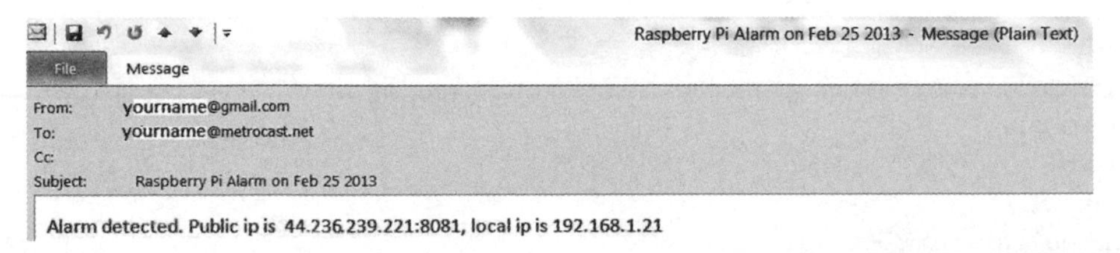

Figure 8-10 Sample e-mail notification.

Laser Trip Program

The laser trip program sends the alarm e-mail when a low level on GPIO pin 23 is detected by the RasPi. The relay contacts shown as points 1 and 2 in Fig. 8-7 will close when the laser beam is interrupted. The contact closure puts a low voltage on pin 23, which is connected to the relay via the Pi Cobbler and a long cable, as shown in Fig. 8-11.

The complete program, named *alarm.py*, is available on www.mhprofessional.com/raspi and is listed on page 112.

Upon receiving the alarm e-mail, you would open a browser and type in the public IP address with the 8081 port, for example:

```
http://44.236.239.221:8081
```

This should take you directly to Motion's webcam server that is streaming the real-time video. What you do next is up to you, as I mentioned earlier.

I haven't quite figured out what to do when an object, such as my cat or dog, decides to sleep in the path of the laser beam. I could set a maximum number of e-mails to be sent, which would be the sensible thing to do, or I could enable the RasPi to sound a loud buzzer, which might be the fun thing to do. I will leave that modification up to you.

Additional Sensors

I am sure you realize that additional sensors may be incorporated into this minimal security system. The software at present is designed to work with wired sensors, so it would be trivial to add additional door and window sensors. All you need to do is to wire them to selected GPIO pins and simply add the activation checks in the main "forever" loop. As mentioned before, up to four webcams can be managed by the Motion software. However, it is doubtful that the RasPi can handle four webcams. I would suggest that the send_alarm() module be modified to take an argument indicating which sensor has been activated. The send_alarm() module could use this argument to modify the e-mail to indicate which protected area has been tripped, for example, showing that a living room window has been opened. You could also incorporate the sound sensor from the camera control project to indicate suspicious sounds, such as glass breaking. There really are many ways to easily expand this security system with little expense.

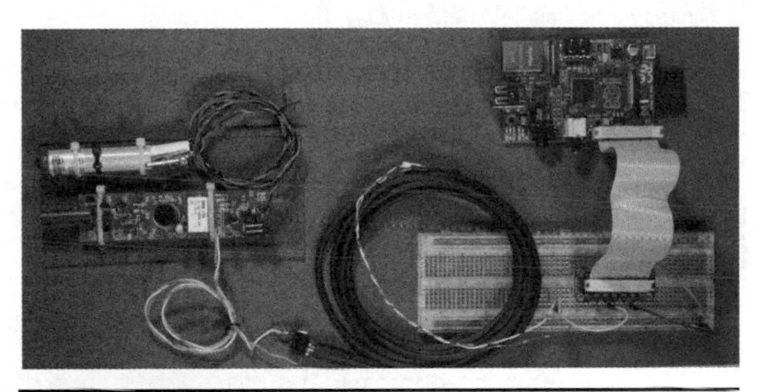

Figure 8-11 Laser trip assembly connected to the RasPi.

You can even include wireless sensors for readers who have built the Z-Wave project in Chap. 7. You would need to explore the open Z-Wave software to see how a software trigger could be created based

alarm.py

```
import subprocess
import smtplib
import socket
from email.mime.text import MIMEText
import datetime
import os
import RPi.GPIO as GPIO
import time
def send_alarm():
    #put the destination e-mail address here
    to = 'youremail@isp.com'
    #put your user gmail account here
    gmail_user = 'user@gmail.com'
    #put your gmail password here
    gmail_password = 'password'
    #everything is verbatim from this line on down
    smtpserver = smtplib.SMTP('smtp.gmail.com', 587)
    smtpserver.ehlo()
    smtpserver.starttls()
    smtpserver.ehlo()
    smtpserver.login(gmail_user, gmail_password)
    today = datetime.date.today()
    arg = 'ip route list'
    proc = subprocess.Popen(arg, shell=True, stdout=subprocess.PIPE)
    data = proc.communicate()
    split_data = data[0].split()
    local_ip = split_data[split_data.index('src') + 1]
    public_ip = os.system('wget http://ipecho.net/plain -O - -q > test.txt; echo')
    public_ip = open('test.txt', 'r').read()
    msg_content = 'Alarm detected. Public ip is %s:8081, local ip is %s' % (public_ip,
    local_ip)
    msg = MIMEText(msg_content)
    msg['Subject'] = 'Raspberry Pi Alarm on %s' % today
    msg['From'] = gmail_user
    msg['To'] = to
    smtpserver.sendmail(gmail_user, [to], msg.as_string())
    smtpserver.quit()
GPIO.setmode(GPIO.BCM)
GPIO.setup(23, GPIO.IN)
# use the count variable in with a 10 sec delay to minimize e-mails sent
count = 0
#begin the forever loop
while True:
    if(GPIO.input(23) == False & count == 0):
        count = count + 1
        send_alarm()
    time.sleep(10)
    count = 0
```

upon a state change on one of the network nodes. I showed a window sensor being activated in Chap. 7, but that state change was only shown in a GUI. I am certain that the underlying logical variable representing that state could be polled and the result sent to the alarm program.

Summary

I started this chapter by discussing the selection of the webcam to be used in the surveillance of the security zone. It turns out that Raspian Wheezy, the RasPi Linux distribution used, has drivers for different manufacturers including many older models.

The laser trip assembly used in the camera controller project was shown in its new role as the primary intrusion sensor. I reused the same interface circuit between the RasPi and the camera controller because it was simple and functional for this project.

The key software package, Motion, was discussed next. I used only a small portion of its capability, that being the webcam server. The Motion setup for the RasPi was discussed along with a demonstration of an actual webcam stream.

I introduced a notification program that will send an e-mail to a designated recipient that contains both the public and local IP addresses currently associated with the RasPi. This program was then incorporated into an overall program that sent the e-mail when the RasPi detected an intrusion signal.

Finally, I concluded with a brief discussion about future modifications and the ease with which additional sensors could be added to expand the system.

CHAPTER 9

NFC/RFID Reader

Introduction

Near Field Communications (NFC) is a contactless communication technology based upon the ISO 14443 standard. Being standards-based, NFC devices can easily interoperate with other contactless protocols including the highly popular *Radio Frequency Identification* (RFID) protocol. RFID tags are often used for building access and asset inventory. In this chapter, I will show you how to interface an NFC/RFID reader with the RasPi. The reader will detect the nearby presence of a tag device and then light an LED.

The reader will be controlled by a software package named *libnfc* and a Python program running on the RasPi. Be forewarned: a fairly involved development process is required to install the libnfc software.

How NFC Works

Magnetic induction is the fundamental physics principle involved with NFC/RFID. Figure 9-1 shows a simplified diagram where the reader, or more formally *Initiator*, emits a radio frequency electromagnetic field that is received by the tag, or *Target*.

The RF field frequency is set at 13.56 MHz, which is in the low-frequency, low-power, Industrial, Service, and Medical (ISM) unlicensed band. While Fig. 9-1 shows a maximum range of 20 cm, a more realistic range is 4 cm or about 1.6 inches. The standard data rates are 106, 212, or 424 kbits/s, which are quite suitable for reader applications and low-to-moderate volume data transfers.

There are two primary modes of operation using NFC/RFID:

1. *Passive*—The Target becomes energized when placed within the Initiator RF field. Only then will it respond with its preprogrammed data.
2. *Active*—Both the Initiator and Target have self-energized RF fields and can establish and maintain communication when in range of each other.

Until recently, the passive mode has been the most popular; however, the use of active mode NFC among smartphone user has become widespread. Smartphone manufacturers are aggressively marketing this new feature to enhance the data sharing capabilities of their products.

NFC Hardware

I will be using the PN532/NFC breakout board manufactured by microBuilder.eu and distributed in the United States by Adafruit Industries, part number 364. Adafruit also sells a similar board, the NFC Shield, part number 789, but this is designed to work with the Arduino microcontroller board

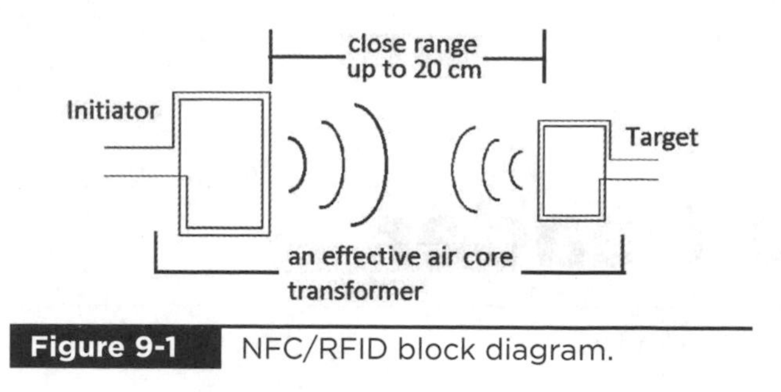

Figure 9-1 NFC/RFID block diagram.

and is *not* the board that will be used in this project. Figure 9-2 shows the breakout board.

The PN532/NFC breakout board has been designed to work with the open-source library libnfc, which will be discussed in this chapter's software section. A full board schematic may be downloaded from the microBuilder website, http://www.microbuilder.eu/Projects/PN532.

There are other interesting links at this website, including a detailed discussion on how the board antenna was designed and optimized for this NFC application. The board uses an NXP PN532 microcontroller (hence its name). The controller is a hybrid type in that it processes both analog RF and digital signals. In Chap. 7, I introduced another hybrid chip, the ZW0301, which processed Z-Wave signals. This chip is similar in operation to the NXP PN532 microcontroller and uses a somewhat familiar ISO-layer network communications model that I discussed in previous chapters.

NFC Data Communications Flow

The information in this section is based upon the discussions and figures contained in the PN532

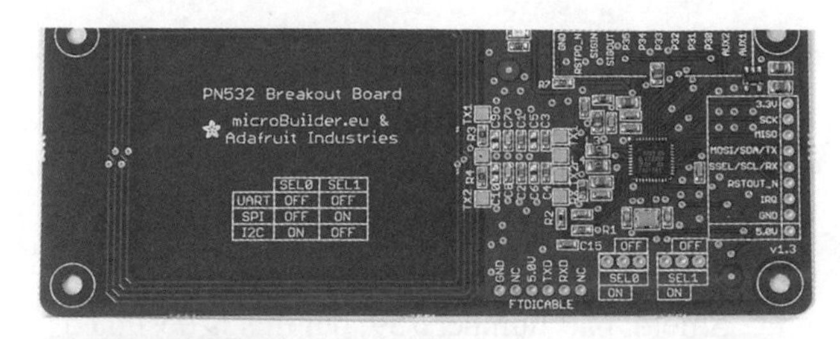

Figure 9-2 PN532/NFC breakout board.

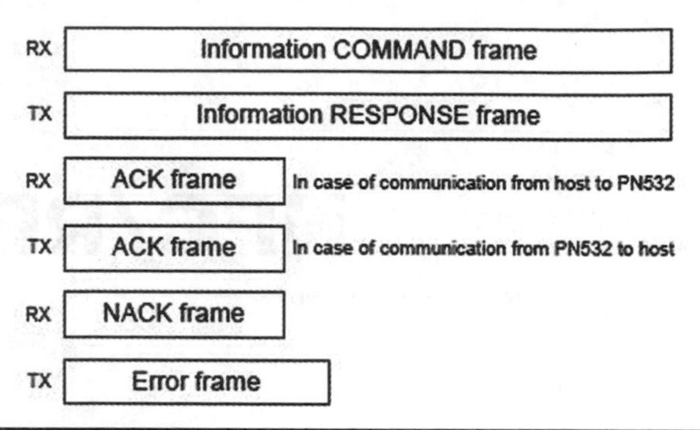

Figure 9-3 Sample data interchange between the RasPi and the PN532 controller.

User's Manual that is available from the Adafruit Industries website. Figure 9-3 shows the format of a sample data interchange between the host computer, which would be the RasPi, and the PN532 controller.

You should notice that the data interchange involves not only sending and receiving commands, but also the interchange of ACKs and NACKs. This is similar to the TCP protocol in that it ensures error-free data communication. Figure 9-4 is a graphical representation of the same data interchange.

The libnfc software package controls all the data communication between the PN532 reader and the RasPi. The reader's firmware will "push" data to the RasPi when it detects the presence of a tag.

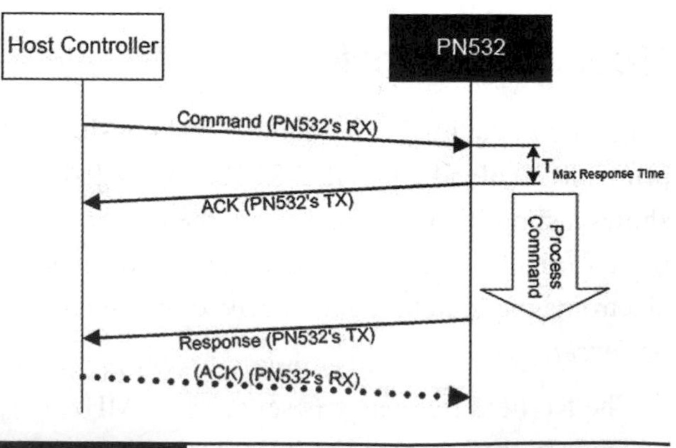

Figure 9-4 Graphical representation of the data interchange between the RasPi and the PN532 controller.

Installing libnfc

This libnfc installation is based upon an excellent tutorial created by Kevin Townsend from Adafruit Industries. I have performed this installation using both the Wheezy 2_2013 and Occidentalis v2.0 distributions. The project Python test code uses the Wheezy distribution.

The PN532 breakout board provides hardware support for UART, SPI, or I²C communication protocols. This project uses the UART protocol because the Wheezy distribution supports it without any modifications as of the time of this writing.

What follows is a step-by-step process that is complemented by many screenshots to help you navigate through this somewhat complex process. I also found it easier to use a SSH connection to accomplish this installation. I refer you to an earlier discussion in Chap. 7 on how to run a SSH session.

The initial steps are to make the Wheezy UART logical connection available to libnfc. The default Wheezy image has it connected to a console named ttyAMA0, which prevents it from being used by another process.

1. Type the following at a terminal command prompt:

   ```
   sudo nano /boot/cmdline.txt ↵
   ```

   Edit the single line of text so that it matches what is shown in Fig. 9-5. All references to ttyAMA0 are removed.

CAUTION Don't change anything else or you will not be able to reboot the RasPi!

2. Type the following at a terminal command prompt:

   ```
   sudo nano /etc/inittab ↵
   ```

   Add a # sign in front of the line below.

   ```
   T0:23:respawn:/sbin/getty –L
   ttyAMA0 115200 vt100.
   ```

   Figure 9-6 shows the edited file.

3. Type the following at a terminal command prompt:

   ```
   sudo reboot ↵
   ```

   A reboot is needed to proceed with the rest of the installation.

4. The source code now needs to be downloaded into a new directory named libnfc. Ensure that you are in the pi directory before starting this step.

   Type the following series of commands:

   ```
   mkdir libnfc ↵
   cd libnfc ↵
   sudo wget http://libnfc.googlecode
   .com/files/libnfc-1.6.0-rc1.tar.gz ↵
   ```

   Figure 9-7 shows a successful download of the libnfc source code.

5. The downloaded source needs to be "unzipped." A new subdirectory named libnfc-1.6.0-rc1 will automatically be created in the libnfc directory after this action is completed.

```
GNU nano 2.2.6                    File: /boot/cmdline.txt

dwc_otg.lpm_enable=0  console=tty1 root=/dev/mmcblk0p2 rootfstype=ext4 elevator=deadline rootwait
```

Figure 9-5 The cmdline.txt edited content.

```
#Spawn a getty on Raspberry Pi serial line
#T0:23:respawn:/sbin/getty -L ttyAMA0 115200 vt100
```

Figure 9-6 The inittab edited content.

Type the following at a terminal command prompt:

```
sudo tar –xvzf libnfc-1.6.0-rc1.
tar.gz ↵
```

Figure 9-8 shows the end portion of the display after this extract operation has finished.

6. One of the source code files must be slightly modified in order for libnfc to connect to the RasPi through the ttyAMA logical device. Type the following at a terminal command prompt:

```
cd libnfc-1.6.0-rc1 ↵
sudo nano libnfc/buses/uart_posix.c
↵
```

You need to add "ttyAMA" to the following line in the file:

```
char *serial_ports_device_radix[] =
{"ttyUSB", "ttyS", NULL};
```

Figure 9-9 shows the file content after the edit has been done.

CAUTION **Ensure that you use double quotes around ttyAMA or the logical device will not be recognized. The strange fact is that the file will compile OK, but the device ttyAMA will never be recognized and the board will never connect. A frustrating experience to say the least.**

7. libnfc must next be configured to operate with the RasPi and the breakout board. Type the following at a terminal command prompt:

```
sudo ./configure --with-drivers-
pn532_uart --enable-serial-
autoprobe ↵
```

Figure 9-10 shows the end results of the configure operation. Just confirm that the line

```
pn532_uart...... yes
```

appears as the last line.

```
pi@raspberrypi ~ $ cd libnfc
pi@raspberrypi ~/libnfc $ sudo wget http://libnfc.googlecode.com/files/libnfc-1.6.0-rc1.t
ar.gz
--2013-03-07 13:04:06--  http://libnfc.googlecode.com/files/libnfc-1.6.0-rc1.tar.gz
Resolving libnfc.googlecode.com (libnfc.googlecode.com)... 173.194.68.82, 2607:f8b0:400d:
c00::52
Connecting to libnfc.googlecode.com (libnfc.googlecode.com)|173.194.68.82|:80... connecte
d.
HTTP request sent, awaiting response... 200 OK
Length: 571772 (558K) [application/x-gzip]
Saving to: `libnfc-1.6.0-rc1.tar.gz.1'

100%[===============================================>] 571,772     1.41M/s   in 0.4s

2013-03-07 13:04:06 (1.41 MB/s) - `libnfc-1.6.0-rc1.tar.gz.1' saved [571772/571772]

pi@raspberrypi ~/libnfc $
```

Figure 9-7 The libnfc source code download.

```
libnfc-1.6.0-rc1/examples/
libnfc-1.6.0-rc1/examples/pn53x-sam.c
libnfc-1.6.0-rc1/examples/nfc-anticol.c
libnfc-1.6.0-rc1/examples/nfc-emulate-tag.c
libnfc-1.6.0-rc1/examples/pn53x-diagnose.1
libnfc-1.6.0-rc1/examples/pn53x-tamashell-scripts/
libnfc-1.6.0-rc1/examples/pn53x-tamashell-scripts/UltraLightRead.cmd
libnfc-1.6.0-rc1/examples/pn53x-tamashell-scripts/ReadMobib.sh
libnfc-1.6.0-rc1/examples/pn53x-tamashell-scripts/ReadNavigo.sh
libnfc-1.6.0-rc1/examples/pn53x-tamashell-scripts/Makefile.in
libnfc-1.6.0-rc1/examples/pn53x-tamashell-scripts/UltraLightReadWrite.cmd
libnfc-1.6.0-rc1/examples/pn53x-tamashell-scripts/Makefile.am
libnfc-1.6.0-rc1/examples/CMakeLists.txt
libnfc-1.6.0-rc1/examples/nfc-poll.1
libnfc-1.6.0-rc1/examples/nfc-dep-initiator.c
libnfc-1.6.0-rc1/examples/doc/
libnfc-1.6.0-rc1/examples/doc/quick_start_example1.c
libnfc-1.6.0-rc1/examples/nfc-relay.c
libnfc-1.6.0-rc1/examples/pn53x-sam.1
libnfc-1.6.0-rc1/examples/nfc-emulate-uid.c
libnfc-1.6.0-rc1/examples/pn53x-tamashell.c
libnfc-1.6.0-rc1/examples/pn53x-diagnose.c
libnfc-1.6.0-rc1/examples/Makefile.in
libnfc-1.6.0-rc1/examples/nfc-emulate-forum-tag2.c
libnfc-1.6.0-rc1/examples/nfc-dep-target.c
libnfc-1.6.0-rc1/examples/nfc-relay.1
libnfc-1.6.0-rc1/examples/nfc-anticol.1
libnfc-1.6.0-rc1/examples/nfc-poll.c
libnfc-1.6.0-rc1/examples/nfc-emulate-tag.1
libnfc-1.6.0-rc1/examples/nfc-dep-target.1
libnfc-1.6.0-rc1/examples/nfc-emulate-uid.1
libnfc-1.6.0-rc1/examples/Makefile.am
libnfc-1.6.0-rc1/examples/nfc-dep-initiator.1
libnfc-1.6.0-rc1/examples/pn53x-tamashell.1
pi@raspberrypi ~/libnfc $
```

Figure 9-8 The libnfc source code extract operation.

8. The preparation for the actual build of the source code starts next. Type the following at a terminal command prompt:

```
sudo make clean ↵
```

This preps all the source code, libraries, etc. for the actual build. Figure 9-11 shows the end of the prep stage.

9. The actual build happens in this stage. I have included two screenshots showing the start and end of the process. This build will take about 30 minutes, so be patient. Type the following at a terminal command prompt:

```
sudo make install all ↵
```

Figure 9-12 shows the start of the build.

Figure 9-13 shows the end of the build. There will now be a complete libnfc installation on the RasPi after this screen appears.

```
char *serial_ports_device_radix[] = { "tty.SLAB_USBtoUART", NULL };
#  elif defined (__FreeBSD__) || defined (__OpenBSD__)
char *serial_ports_device_radix[] = { "cuaU", "cuau", NULL };
#  elif defined (__linux__)
char *serial_ports_device_radix[] = { "ttyUSB", "ttyS", "ttyAMA", NULL };
```

Figure 9-9 The uart_posix.c source code edit.

```
configure: creating ./config.status
config.status: creating Doxyfile
config.status: creating Makefile
config.status: creating cmake/Makefile
config.status: creating cmake/modules/Makefile
config.status: creating contrib/Makefile
config.status: creating contrib/devd/Makefile
config.status: creating contrib/udev/Makefile
config.status: creating contrib/win32/Makefile
config.status: creating contrib/win32/sys/Makefile
config.status: creating examples/Makefile
config.status: creating examples/pn53x-tamashell-scripts/Makefile
config.status: creating include/Makefile
config.status: creating include/nfc/Makefile
config.status: creating libnfc.pc
config.status: creating libnfc/Makefile
config.status: creating libnfc/buses/Makefile
config.status: creating libnfc/chips/Makefile
config.status: creating libnfc/drivers/Makefile
config.status: creating test/Makefile
config.status: creating utils/Makefile
config.status: creating config.h
config.status: config.h is unchanged
config.status: executing depfiles commands
config.status: executing libtool commands

Selected drivers:
   acr122........... no
   acr122s.......... no
   arygon........... no
   pn53x_usb........ no
   pn532_uart....... yes
pi@raspberrypi ~/libnfc/libnfc-1.6.0-rc1 $
```

Figure 9-10 End of the configure operation.

```
make[2]: Leaving directory `/home/pi/libnfc/libnfc-1.6.0-rc1/libnfc/buses'
Making clean in chips
make[2]: Entering directory `/home/pi/libnfc/libnfc-1.6.0-rc1/libnfc/chips'
rm -rf .libs _libs
test -z "libnfcchips.la" || rm -f libnfcchips.la
rm -f "./so_locations"
rm -f *.o
rm -f *.lo
make[2]: Leaving directory `/home/pi/libnfc/libnfc-1.6.0-rc1/libnfc/chips'
Making clean in .
make[2]: Entering directory `/home/pi/libnfc/libnfc-1.6.0-rc1/libnfc'
test -z "libnfc.la" || rm -f libnfc.la
rm -f "./so_locations"
rm -rf .libs _libs
rm -f *.o
rm -f *.lo
make[2]: Leaving directory `/home/pi/libnfc/libnfc-1.6.0-rc1/libnfc'
make[1]: Leaving directory `/home/pi/libnfc/libnfc-1.6.0-rc1/libnfc'
Making clean in .
make[1]: Entering directory `/home/pi/libnfc/libnfc-1.6.0-rc1'
test -z "Doxygen.log coverage.info libnfc.pc" || rm -f Doxygen.log coverage.info libnfc.pc
rm -rf .libs _libs
rm -rf doc
rm -rf coverage
rm -f *.lo
make[1]: Leaving directory `/home/pi/libnfc/libnfc-1.6.0-rc1'
pi@raspberrypi ~/libnfc/libnfc-1.6.0-rc1 $
```

Figure 9-11 The end of the build prep.

```
pi@raspberrypi ~/libnfc/libnfc-1.6.0-rc1 $ sudo make install all
Making install in libnfc
make[1]: Entering directory `/home/pi/libnfc/libnfc-1.6.0-rc1/libnfc'
Making install in chips
make[2]: Entering directory `/home/pi/libnfc/libnfc-1.6.0-rc1/libnfc/chips'
  CC     libnfcchips_la-pn53x.lo
```

Figure 9-12 The Build start.

```
make[2]: Entering directory `/home/pi/libnfc/libnfc-1.6.0-rc1/test'
make[2]: Nothing to be done for `all'.
make[2]: Leaving directory `/home/pi/libnfc/libnfc-1.6.0-rc1/test'
make[2]: Entering directory `/home/pi/libnfc/libnfc-1.6.0-rc1'
make[2]: Leaving directory `/home/pi/libnfc/libnfc-1.6.0-rc1'
make[1]: Leaving directory `/home/pi/libnfc/libnfc-1.6.0-rc1'
pi@raspberrypi ~/libnfc/libnfc-1.6.0-rc1 $
```

Figure 9-13 The build end.

Hardware Installation

The interface between the RasPi and the PN532 breakout board requires only four wires, as shown in the block diagram in Fig. 9-14. The Pi Cobbler prototype tool is also used to ease this installation.

I also connected an LED to pin 23 with a series-current-limiting resistor. The LED will provide a visual indication when a tag is detected, using the Python program. Figure 9-15 is a picture of the physical layout of the hardware.

Do not be concerned with the fact that 5 V is being connected to the PN532 board, while there are direct connections to the UART lines, TXD and RXD. The PN532 shifts the voltage such that only 3.3 V appears on the UART lines, thus protecting the RasPi from any overvoltage being applied to the GPIO pins.

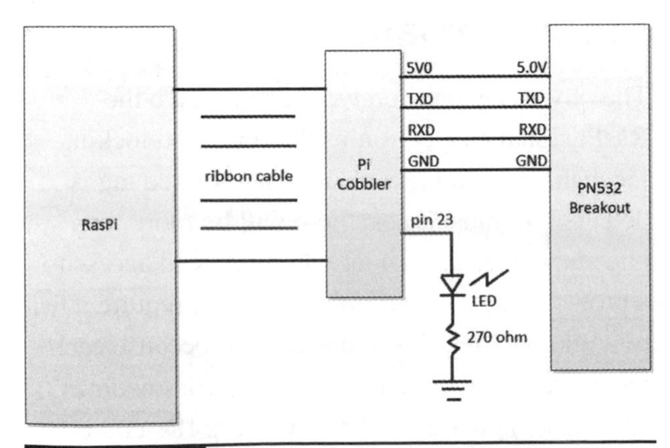

Figure 9-14 The RasPi and the PN532 interface.

Figure 9-15 Project hardware layout.

Initial Checkout

There is a very useful test program named *nfc-poll*, which is provided in the libnfc package and will test the hardware and software for proper operation. First, ensure that you are in this directory:

```
/home/pi/libnfc/libnfc-1.6.0-rc1
```

Then enter the following at the command line prompt:

```
sudo examples/nfc-poll ↵
```

Put a tag on or near the reader, and you should see something similar to the Fig. 9-16 screenshot.

Of course, your UID tag information will be different from what is shown in the figure, but everything else should be the same. Successful program execution means that you are now ready for the project program.

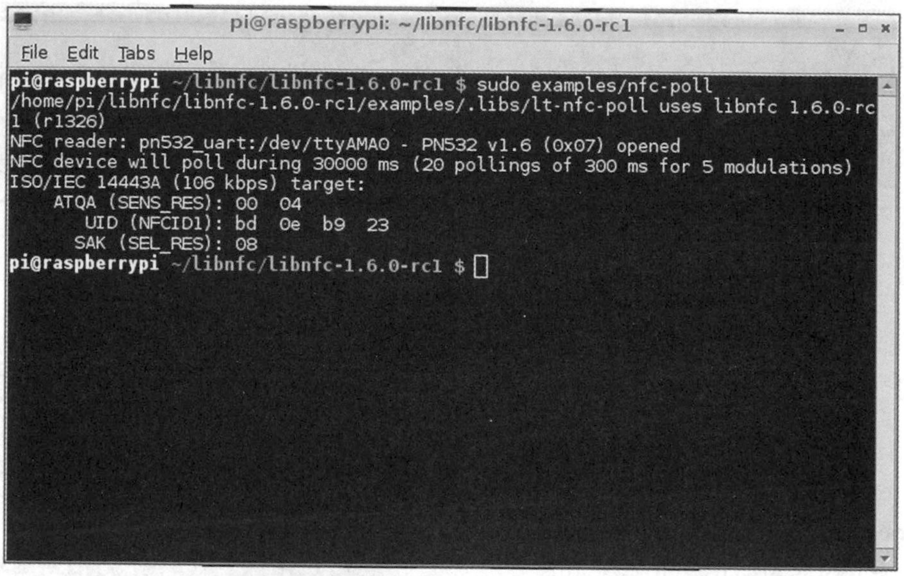

```
pi@raspberrypi: ~/libnfc/libnfc-1.6.0-rc1
File  Edit  Tabs  Help
pi@raspberrypi ~/libnfc/libnfc-1.6.0-rc1 $ sudo examples/nfc-poll
/home/pi/libnfc/libnfc-1.6.0-rc1/examples/.libs/lt-nfc-poll uses libnfc 1.6.0-rc
1 (r1326)
NFC reader: pn532_uart:/dev/ttyAMA0 - PN532 v1.6 (0x07) opened
NFC device will poll during 30000 ms (20 pollings of 300 ms for 5 modulations)
ISO/IEC 14443A (106 kbps) target:
    ATQA (SENS_RES): 00  04
       UID (NFCID1): bd  0e  b9  23
      SAK (SEL_RES): 08
pi@raspberrypi ~/libnfc/libnfc-1.6.0-rc1 $
```

Figure 9-16 The nfc-poll test program output.

Project Program

The project program is designed to check for a predesignated tag ID and then turn on an LED if the tag is detected by the reader. This program provides the basic functionality of the NFC design and serves as a template to expand the application for additional uses, which will be described later.

The Python code listing is shown on page 123. This program named extract_data.py uses the nfc-poll program to capture the tag information and writes it to a text file named nfc_data.txt. This file is then read and parsed to extract the tag ID data. This ID data is then compared to a hard-coded tag ID to check if it matches. A match will turn on the LED for 10 seconds as well as display the word "match" on the console screen. If the tag doesn't match, the phrase "no match" appears on the console.

Run the program from the pi directory by entering the following at the command prompt:

```
sudo python extract_data.py ↵
```

The console display cursor will jump to a new line and wait for a tag to be placed near the reader. Once this is done, the display will show "match" or "no match," and the LED will light for 10 seconds only on a match condition.

This program is quite constrained in that it must have the desired tag ID hard coded before it is run. But as previously stated, it is only a starter or template that you can use to greatly increase the flexibility and utility of the project. Suggested modifications and improvements are discussed in the next section.

Future Expansion

The obvious expansion would be to have the RasPi signal an electronically controlled locking mechanism to unlock when an authorized tag is detected. In most cases, there will be more than one authorized tag holder who will need access to a protected area or building. This will require a list of authorized tag IDs to be checked because each tag contains a unique ID. Python accommodates several ways to do this function. Tag IDs could even be placed in a database that the program could

extract_data.py

```
import time
import os
import RPi.GPIO as GPIO
GPIO.setwarnings(False)
GPIO.setmode(GPIO.BCM)
GPIO.setup(23, GPIO.OUT)
nfc_id = None
while (nfc_id == None):
    nfc_id = os.system('sudo libnfc/libnfc-1.6.0-rc1/examples/nfc-poll > nfc_data.
    txt')
inFile = open('nfc_data.txt')
lines = inFile.readlines()
inFile.close()
buffer = []
for line in lines:
    line_content = line.split()
    if(not line_content[0] == 'UID'):
        pass
    else:
        buffer.append(line_content)
#print(buffer)
str = buffer[0]
id_str = str[2] + str[3] + str[4] + str[5]
#print(id_str)
if(id_str == 'bd0eb923'):
    print('match')
    GPIO.output(23, GPIO.HIGH)
    time.sleep(10)
    GPIO.output(23, GPIO.LOW)
else:
    print('no match')
```

query. This is actually the way such an access function is accomplished in real-world situations that use commercial systems. Such systems are also very expensive.

Another expansion might be to have the program send a preset e-mail when a specific tag is detected. The e-mail program shown in Chap. 8 could easily be incorporated into this project program. The e-mail would be automatically created and sent when a tag is detected. I could envision a situation in which a child or perhaps a disabled person might need to send a non-urgent e-mail for help or assistance. Obviously, calling 911 (or the international equivalent) should be used in an emergency. All the person need do is to swipe the tag near the reader, and the e-mail will be sent. An LED should also light to confirm that the e-mail was sent.

Other applications that might use tag-initiated activities include home tasks, such as starting an irrigation system, pool cleaning, garage door opening/closing, home spa operations, and so on.

Summary

This chapter began with an explanation of what NFC is and how it works. The two primary operational modes were shown.

The PN532 breakout board was discussed next, along with the rationale as to why it was selected for this project.

This was followed by a brief overview of the NFC communication interchange, which illustrated that it was a robust, TCP-like data flow between the RasPi and the PN532.

A comprehensive, nine-step procedure was detailed, showing how to install the open-source libnfc software package. This package is installed in the Wheezy Linux distribution run on the RasPi.

A relatively easy hardware install was covered along with a quick test program to prove that the PN532 could detect a tag.

The project program was discussed next and then executed, causing a preset tag ID to be detected and an LED to be lit to show that the tag had been detected.

The chapter concluded with suggestions for further improvements, modifications, and expansions.

CHAPTER 10

1-Wire Weather Station

Introduction

As the chapter's name implies, I will be showing you how to build a weather station. This station will be quite basic, measuring only temperature, pressure, and humidity, but it may be easily expanded to measure other weather components, such as wind speed and rainfall. All the weather sensors are connected to the RasPi with a very clever interface known as *1-Wire*™, which makes it very easy to add additional components.

The RasPi runs on a software package named *owfs*, short for 1-Wire File System. This open-source software contains every function required to interface the RasPi with all the system components used in this project.

Weather Station Design

Figure 10-1 is a system block diagram that shows the components or modules involved in this project.

The system is a point-to-point interconnection, starting with the USB to 1-Wire adapter that is plugged into a port on a powered USB hub, which, in turn, is connected to one of the RasPi USB ports. I didn't show the hub on the diagram because it should be considered as part of the basic RasPi configuration.

All the modules and the USB adapter are available in the United States from www.hobby-boards.com. Similar modules are available from www.sheepwalkelectronics.co.uk for international readers. These items are listed in Table 10-1.

The first element shown connected to the RasPi in the system diagram is an adapter in the physical form of a USB dongle, as shown in Fig. 10-2. It has a USB connector at one end and a 1-Wire connector at the opposite side.

The 1-Wire socket shown in the figure is a standard RJ12, six-pin, telephone-style connector. A special RJ12 to RJ45 cable is required to go from the dongle to the first 1-Wire module, which

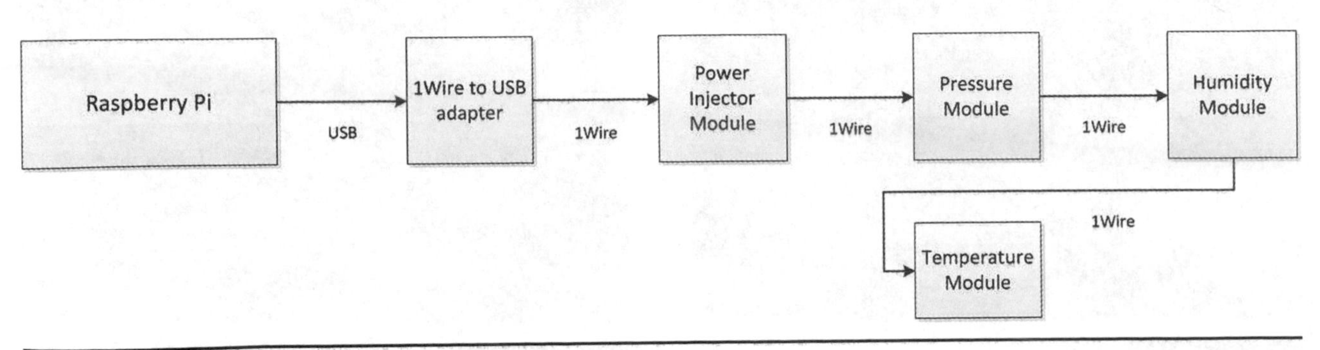

Figure 10-1 Weather station block diagram.

Table 10-1 Listing of Weather Station Modules and Adapters

Module Name	Hobby Board	Sheepwalk Electronics	Remarks
USB to 1-Wire Adapter	DS9490R-P	DS9490	
Temperature	T4-R1-A	SWE1	
Barometer	B1-R1-A		Measures pressure
Humidity	H3-R1-A	SWE3	
Power Injector	PI2-R1-A	SWE5	
1-Wire Sniffer	SN1-R1-A		Not required for weather station, but helpful to debug and program.

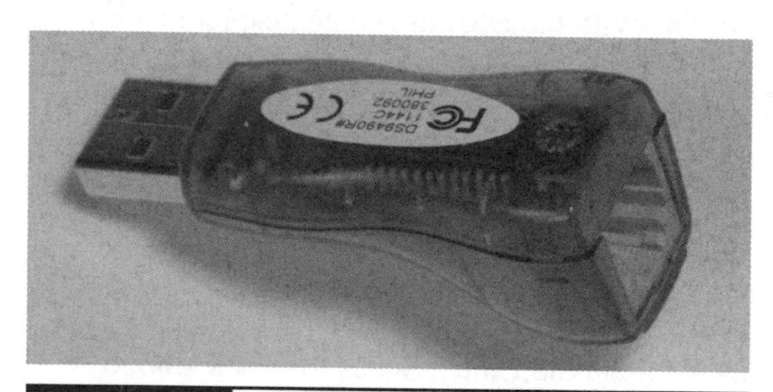

Figure 10-2 USB to 1-Wire dongle.

in this case, is the Power Injector Module. This special cable is shown in Fig. 10-3 and is typically provided when you purchase the 1-Wire-to-USB dongle.

Most 1-Wire modules use the standard eight-pin RJ45 connection standard, which is exactly the same as a regular Ethernet patch cable. In fact, I used normal Ethernet patch cables to interconnect all the 1-Wire modules for this project.

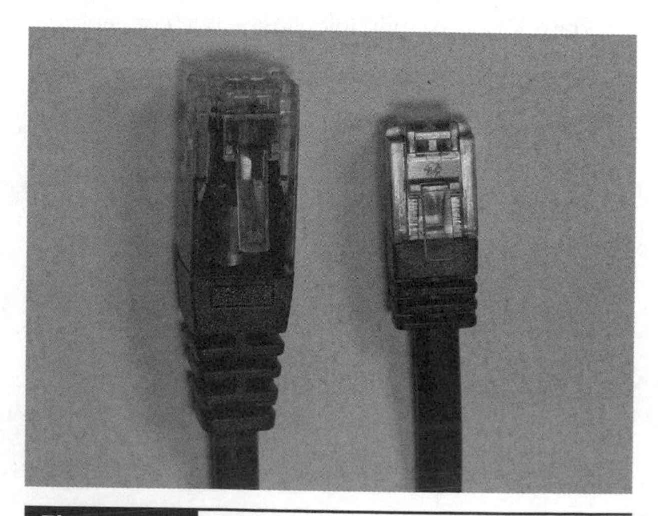

Figure 10-3 RJ12 to RJ45 cable.

CAUTION **Be sure that you do not inadvertently use an Ethernet crossover cable because that will break the 1-Wire connection scheme. It will not harm the modules, but the data connection will not work.**

The next element shown in the system diagram is the Power Injector Module, which is shown in Fig. 10-4.

A 15-V "wall wart" power supply (a box-shaped device that plugs into the AC electrical wall outlet and provides DC power to various electronic devices) plugs into a socket mounted on the case

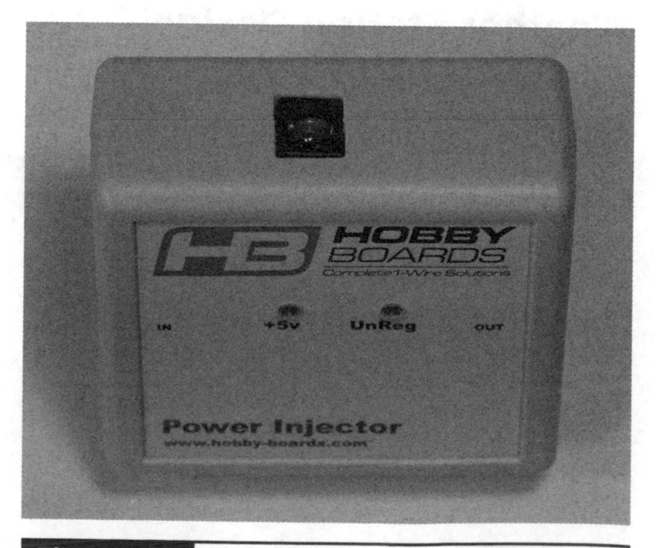

Figure 10-4 Power Injector Module.

top. There are also two RJ45 connectors mounted on the sides of the case that are not visible. The cable from the USB dongle connects to the left socket, while the patch cord that goes to the next module connects to the right socket. The 15 V are sent out through the Ethernet cable and are also down regulated to 5 V for local use by a 78M05 regulator mounted on the internal PCB that is shown in Fig. 10-5.

It turns out that some 1-Wire modules do not require a separate power supply. This is discussed in the section describing various ways to power a 1-Wire network. In this weather station project, the barometer, humidity, and packet sniffer modules do require the use of a power injector because their power consumption exceeds the 1-Wire standards for parasitic power supply operation, which will be discussed shortly.

The weather modules following the Power Injector Module do not have to be inserted in any particular order. I will show you the Temperature Module next. Figure 10-6 shows the external case.

The external case is the same size as the Power Injector Module except that there are two RJ45 sockets at the top but no separate power socket. The RJ45 sockets are wired in parallel, so there

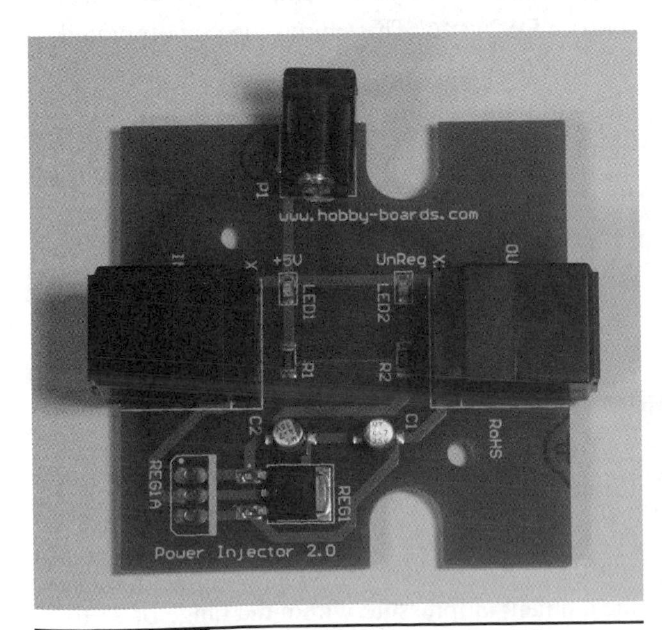

Figure 10-5 Power Injector PCB.

Figure 10-6 Temperature Module case.

are no specific "in" or "out" connections to worry about. The bottom case panel has six holes in it to allow for airflow so that the sensor can measure the ambient air temperature, not just the air contained in the case.

Figure 10-7 shows the PCB inside the Temperature Module.

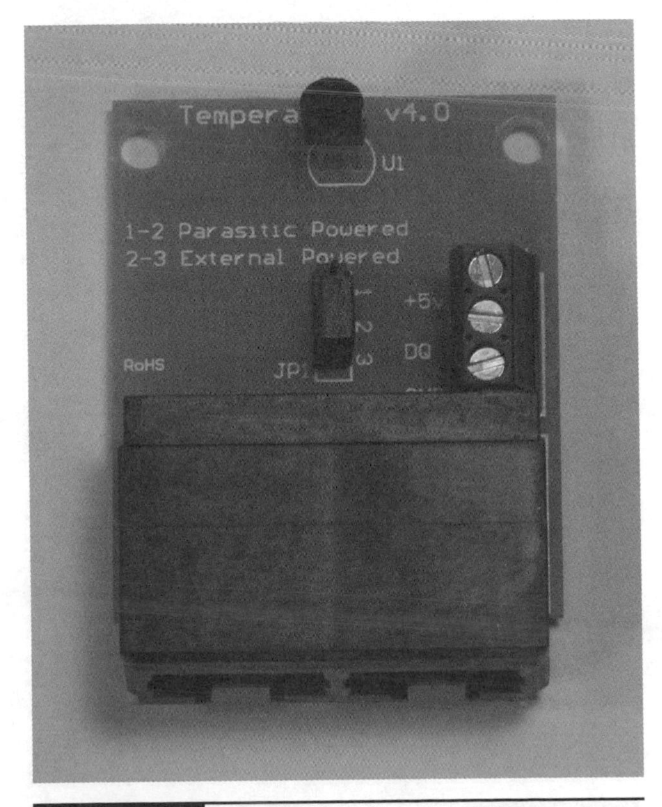

Figure 10-7 Temperature Module PCB.

The temperature sensor, marked U1, may be clearly seen mounted near the top centerline of the board. It appears as if it is an ordinary TO92, a three-lead, plastic-encased transistor, but it is remarkably far more than that. It actually houses a microprocessor with a temperature-sensing element, which is only made possible due to the nature of the 1-Wire protocol. U1 is a Maxim DS18B20 chip, which will be further described later. You should also notice that a three-terminal strip, located at the midpoint of the right side, allows for a separate power supply to be attached and for the module's digital output to be accessed via the terminal marked DQ. We do not use this terminal strip, as all modules are interconnected via the RJ45 cables. It just allows for additional flexibility.

The Humidity Module will be described next. I am not showing an external case picture because it is identical to that of the Temperature Module except for the Humidity designation on the case front. The internal PCB, however, is considerably different, as you may see from Fig. 10-8.

Figure 10-8 Humidity Module PCB.

Figure 10-9 Barometer Module PCB.

The humidity sensor, designated as U3 (not visible in Fig. 10-8), is located at the top of the board. It is a Honeywell model HIH-4021-001 sensor that produces an analog output based on the humidity it senses in the ambient air. These analog sensor readings are sent to chip U1, which is a Maxim DS2438A. This chip is also incorporated into the Barometer Module and will be discussed in a following section. The external terminal connection strip is also present and can be seen on the right-hand side of the board. This strip is not used here, as I mentioned in the Temperature Module discussion.

The Barometer Module is the last sensor to be discussed. Again, no external case picture is needed, but the PCB board is shown in Fig. 10-9.

The pressure sensor, designated as U2, is located on the left side at the top of the board. It is a Freescale Semiconductor model MPXA4115A sensor, which produces an analog output based on the ambient air pressure within the range of 28 to 32 inches of mercury. These readings are sent to

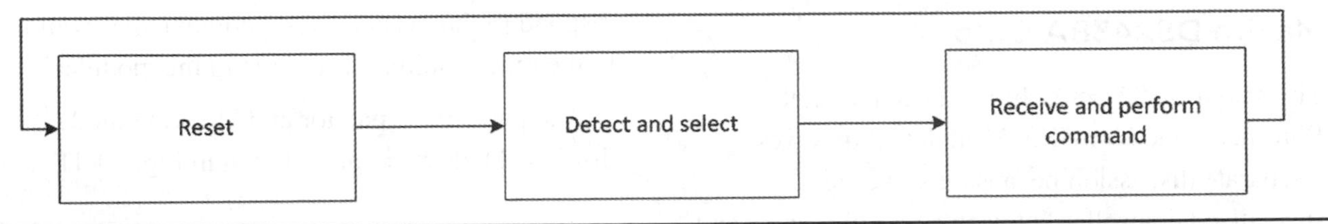

Figure 10-10 1-Wire communication flow.

chip U1, a Maxim DS2438A, the same chip that was used in the Humidity Module. The U1 chip is not visible in this figure because it is surface mounted to the backside of the board. The twice-mentioned terminal board is also visible. A potentiometer is located on the upper right side to precisely calibrate the module to a known air pressure.

I will discuss the Packet Sniffer Module in a later section. Next, I will discuss the 1-Wire protocol, which is the base technology for this project.

1-Wire Protocol

The 1-Wire protocol is the registered trademark name given to a digital serial communications protocol by Dallas Semiconductor, the company that first introduced it in the early 1990s. However, the trademark does not affect any software implementing the protocol. Dallas Semiconductor was merged with Maxim years later but still manufactures 1-Wire products along with many other components. This is the reason why 1-Wire products have a DS prefix.

1-Wire uses a master/slave, multi-drop architecture with an open-drain connection with pull-ups to 5 V. This means that devices may be added or removed without any change to the hardware configuration being specified. All devices are discovered through software techniques. Also all 1-Wire devices have a unique 64-bit identification number that is encoded in a ROM by the manufacturer.

The master and all slaves act as transceivers meaning that they can both transmit and receive, but not at the same time. This mode of operation is known as half-duplex with data transmission being unidirectional. The master initiates all communication on the bus with the slaves responding only to commands sent by the master. All data is sent by serial bits in a specific timed sequence. Bit timing is asynchronous with no external clock required, since all timing is based on the signal transitions from the master.

All communication flow between the master and slaves involves three activities, as shown in Fig. 10-10.

1. *Reset*—In the first phase, the bus master must issue a reset command that synchronizes all elements on the 1-Wire bus. All slaves must respond to the reset, or the bus will not function as desired.
2. *Detect and Select*—A specific slave device is next selected to receive commands from the master in the next phase. This selection is a multipart process that is started by using a binary search algorithm to discover all the slaves currently connected on the bus. Remember, all 1-Wire devices have a unique serial ID permanently programmed into their onboard ROM. The search algorithm reads and records all the IDs and records these values in a dynamic table that is hosted in the master. The master can then use a specific ID to send commands to a slave with that ID, while all other slaves ignore the command.
3. *Receive and Perform Command*—In the last phase, the master and selected slave engage in half-duplex communication in which the master issues commands and the slave responds as designed.

Maxim DS2438A Chip

The Maxim DS2438A chip is used in both the Barometer and Humidity Modules. It deserves a separate discussion because it serves as a specialized controller between a module sensor and the 1-Wire bus. This chip is formally designated as a Smart Battery Monitor; however, it is used in another role for the aforementioned modules. Figure 10-11 is a block diagram illustrating the many functions implemented by this small chip, which is only an eight-lead *small-outline integrated circuit* (SOIC).

The chip contains an analog-to-digital converter (ADC) that converts the sensor's analog signals into equivalent digital signals that can be passed over the 1-Wire network.

The module's unique 64-bit ID is also stored in a ROM contained in this chip. The ROM contains all the programming code needed to respond to the master's reset command as well as to the other commands unique to the module.

The parasitic capacitor and blocking diode discussed below are also shown in Fig. 10-11.

Maxim DS18B20 Chip

This chip is called a Programmable Resolution 1-Wire Digital Thermometer by the manufacturer. Some key specifications are listed below:

- Configurable from 9 to 12 bits
- Measures −55° to +125°C
- Measures −10° to +85°C with a +/− 0.5°C accuracy
- Has its own 64 bit ID
- Powered from 3 to 5.5 V (may be parasitically powered)
- Converts 12 bits within 750 ms

DS2438 BLOCK DIAGRAM

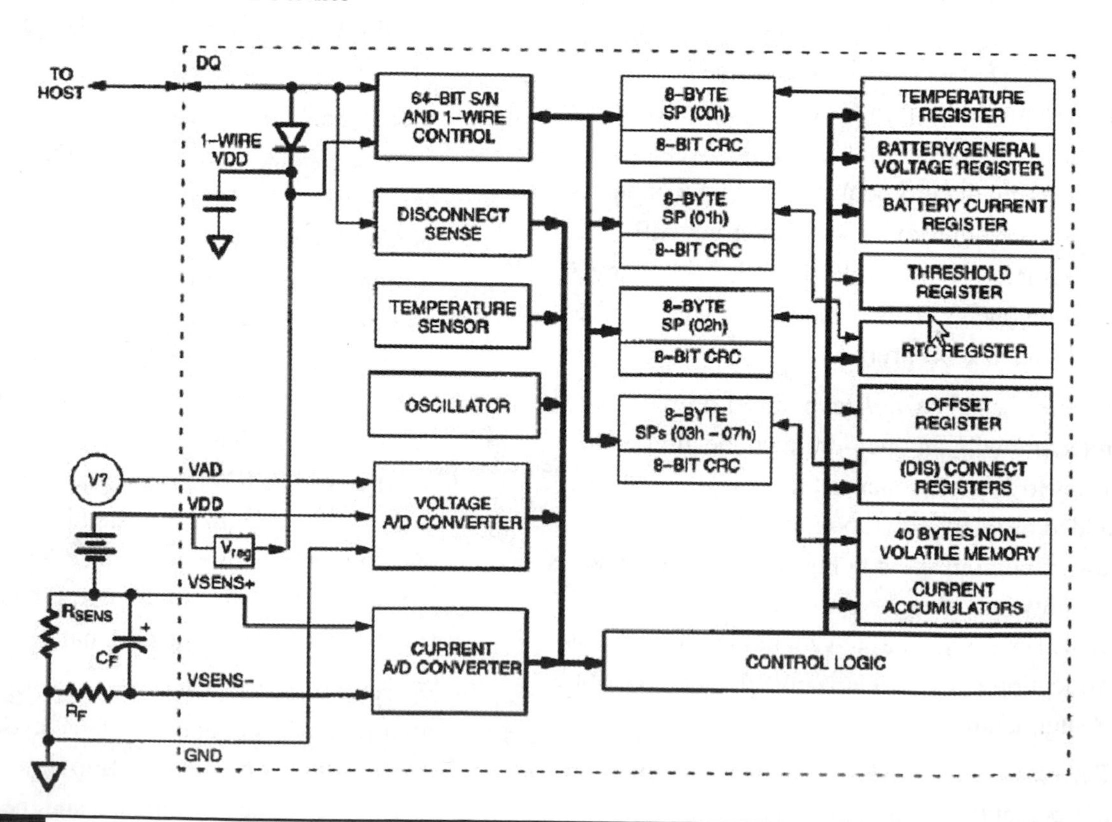

Figure 10-11 DS2438A block diagram.

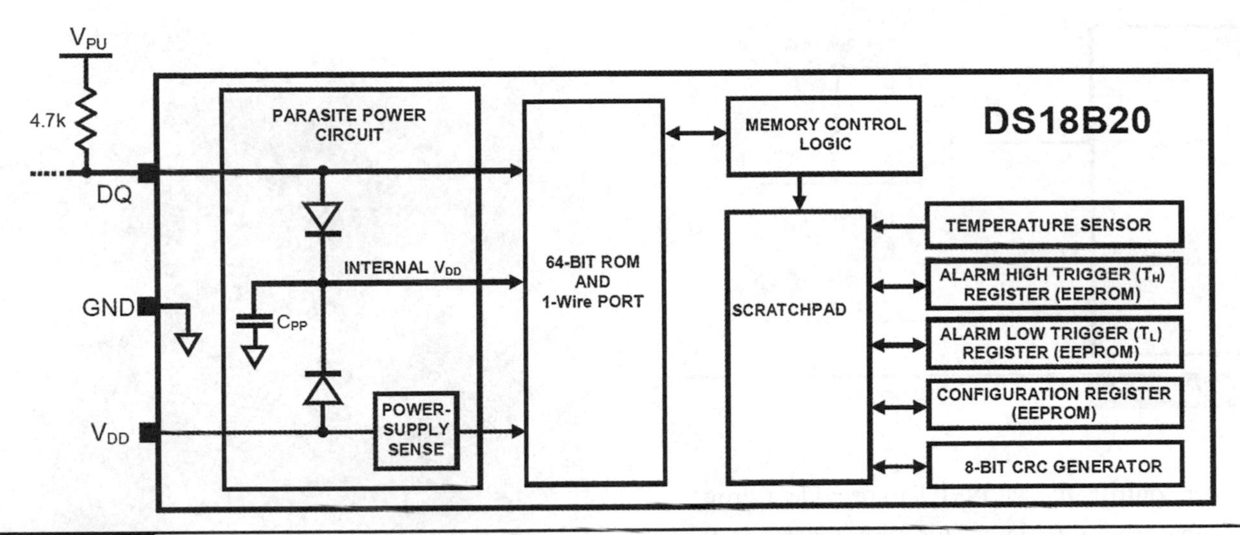

Figure 10-12 DS18B20 block diagram.

Figure 10-12 shows the DS18B20 block diagram. There is an awful lot crammed into a TO92 case.

Powering the 1-Wire Bus

There are three ways that the 1-Wire bus can be powered. These are:

1. External power supply
2. Parasitic supply using a capacitor and diode
3. Powering the data line using a strong pull-up *metal–oxide–semiconductor field-effect transistor* (MOSFET)

The first way, using an external supply, is the one we used in this project by incorporating the Power Injector Module. You were also shown how to connect a local power supply to a specific module using a terminal board located on the PCB. It should be noted that the Power Injector Module is wired to provide power to all bus devices, while a local power supply powers only the device to which it is wired.

The second way is called *parasitic* because the module power is obtained by charging a capacitor from the 1-Wire data line while it is idling at a 5-V level. Figure 10-13 shows the simple charging circuit.

The blocking diode shown in the figure prevents the charged capacitor from interfering with the voltage levels on the data line. There are two disadvantages to using this scheme:

1. The increased capacitive load on the bus data line decreases the effective physical length of the bus line.
2. The capacitor charge can supply only a very small current before being discharged.

The third way, using an active device such as a MOSFET, allows for power to be supplied to the data line during idle time. It essentially improves the parasitic power supply performance because the

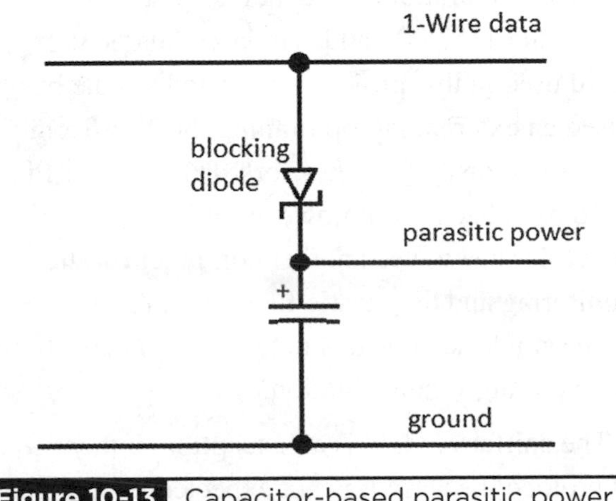

Figure 10-13 Capacitor-based parasitic power supply.

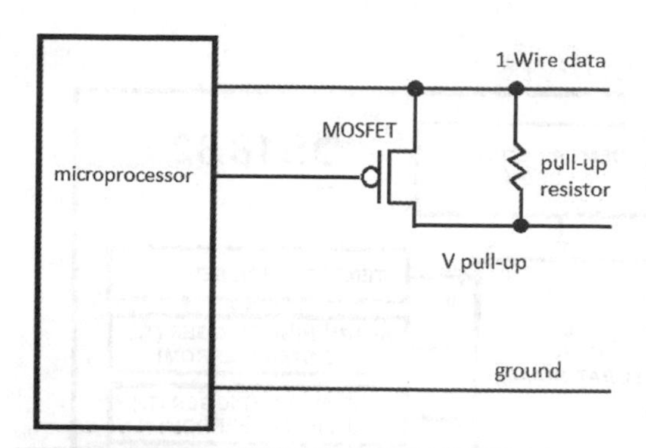

Figure 10-14 Active pull-up power supply.

master controls the MOSFET to provide a strong pull up during idle time. This circuit is shown in Fig. 10-14.

The active pull up provides for extended range and a much better current supply at the expense of requiring a dedicated microprocessor GPIO pin as well as additional programming for that GPIO pin.

1-Wire Sniffer

The 1-Wire Sniffer is a diagnostic module that connects to the 1-Wire network and provides a real-time view of the data flow. It has an RS-232 serial port that may be seen at the top of the PCB, as shown in Fig. 10-15.

The RS-232 serial connection can be attached to any computer running a terminal control program with a configuration of 8 data bits, 1 stop bit, no parity, and 115,200 baud. The particular sniffer board used in this project is powered via the bus. I used an external laptop running the Tera Term program to display the network data. The RasPi could have been used; however, it was simpler and easier just to use another computer for the monitoring and diagnostic functions. You will see some sample screenshots in the section describing the operating weather station.

The sniffer module is an exception to the unique ID protocol because it is not expected to operate as a normal slave device. All of the sniffers made

Figure 10-15 1-Wire Sniffer PCB.

by this manufacturer have the same ID of 0x01, which means that only one can be deployed into a given 1-Wire network. That's usually not an issue because you would never require more than one sniffer per network. The master can issue a limited number of commands to this sniffer, including:

- F—Puts Sniffer into firmware upgrade mode
- S—Puts Sniffer into the sniffer mode
- P—Pauses the Sniffer
- H—Display help message

Set Up the Weather Station Hardware

The weather station modules should now be interconnected using Ethernet patch cables, and the USB dongle should be plugged into a powered USB hub. Plug the cable from the USB dongle into the Power Injector Module. All the other modules are daisy-chained from the Power Injector Module. The 15 V wall wart power supply must also be plugged into the Power Injector Module. Figure 10-16 shows the whole setup.

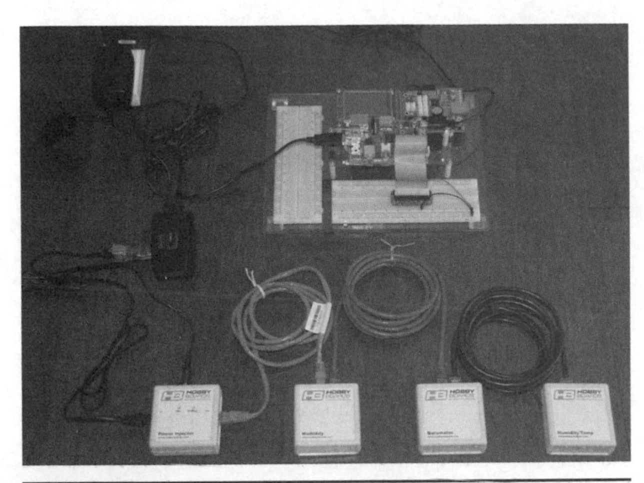

Figure 10-16 The 1-Wire weather station.

1-Wire File System

The software package that drives the whole 1-Wire network is named owfs, as mentioned in the introduction. It is an open-source, comprehensive set of applications that works seamlessly with the RasPi Wheezy distribution. The following are the instructions on how to set up the RasPi to run owfs:

1. Update and upgrade the Wheezy distribution:

   ```
   sudo apt-get update ↵
   sudo apt-get upgrade ↵
   ```

   (this might take some time).

2. Next download and install owfs:

   ```
   sudo apt-get install owfs ↵
   ```

3. Next, edit the owfs configuration file.

   ```
   sudo nano /etc/owfs.conf ↵
   ```

   There are two changes to make. Refer to Fig. 10-17 to see these changes.

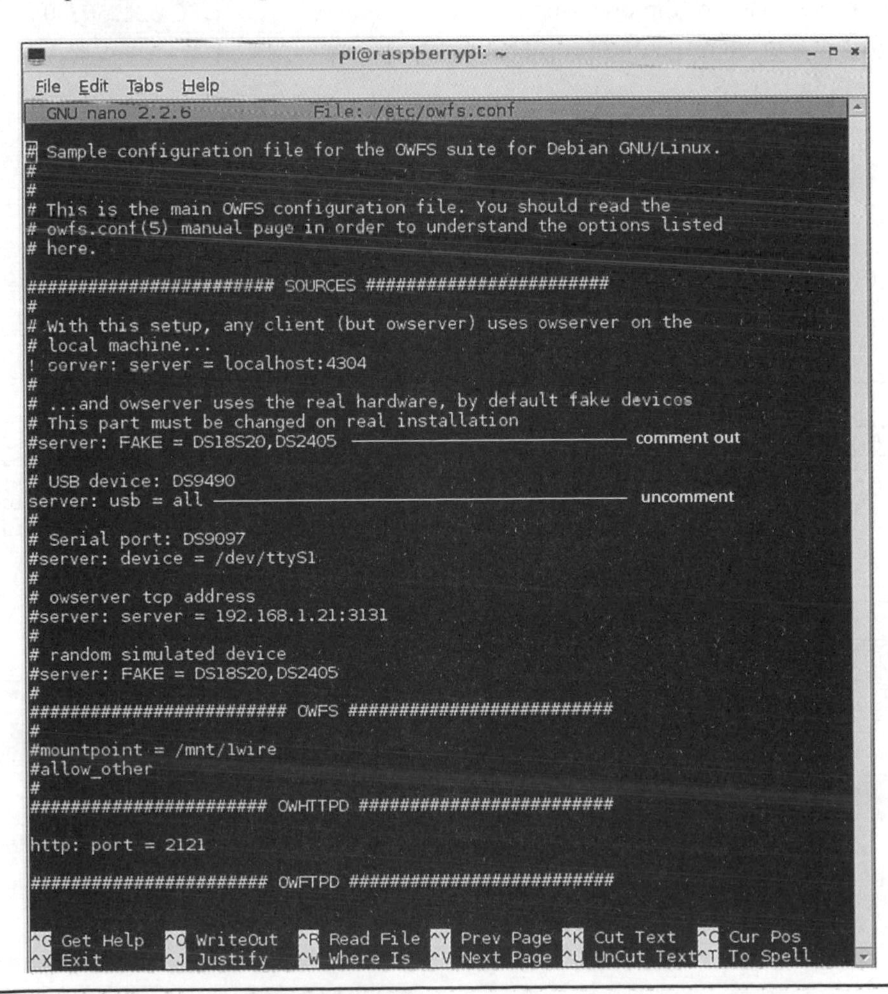

Figure 10-17 The owfs.conf file changes.

4. The last step should be to reboot the RasPi in order to put the configuration changes into effect:

```
sudo reboot ↵
```

Viewing the Weather Data

The real-time data from the modules may be viewed from the owfs server. Open a browser on another computer other than the RasPi, and go to the local RasPi IP address with port 2121 appended. In my case it was:

```
192.168.1.21:2121
```

Figure 10-18 is a screenshot of the owfs server opening page, showing a directory name for each of the modules from Fig. 10-16.

OWFS on 1:6	Bus listing

directory

top	highest level	directory
28.C1EA25030000	28.C1EA25030000	directory
26.882728010000	26.882728010000	directory
26.827BBC000000	26.827BBC000000	directory
81.BA0731000000	81.BA0731000000	directory
bus.0	bus.0	directory
uncached	uncached	directory
settings	settings	directory
system	system	directory
statistics	statistics	directory
structure	structure	directory
simultaneous	simultaneous	directory
alarm	alarm	directory

Figure 10-18 The owfs web server opening page.

The first three directory names listed in the owfs server opening page are for the actual sensor modules, while the fourth is for the USB dongle. The rather odd directory name is how owfs interprets the module ID. For example, by clicking on the directory named 28.C1EA25030000, you open another page, shown in Fig. 10-19, that provides data from the Temperature Module.

A lot of information appears, but the key data is the temperature itself, which is shown at 20.8125 °C. You need to refer to the owfs user's manual to

28.C1EA25030000

uncached version

up	directory
address	28C1EA25030000C4
alias	CHANGE
crc8	C4
errata	errata
family	28
fasttemp	21
id	C1EA25030000
locator	FFFFFFFFFFFFFFFF
power	YES (1)
r_address	C400000325EAC128
r_id	00000325EAC1
r_locator	FFFFFFFFFFFFFFFF
temperature	20.8125
temperature10	20.75
temperature11	20.875
temperature12	20.8125
temperature9	21
temphigh	75 CHANGE
templow	70 CHANGE
type	DS18B20

Figure 10-19 Temperature Module results.

determine what all the other data means except the *temphigh* and *templow* entries, which are self-explanatory. Apparently, those readings are shown in degrees Fahrenheit due to the owfs default configuration setup.

Clicking on the directory named 26.882728010000 opens the Humidity Module page. Figure 10-20 shows the data on this page.

Again, there is plenty of data shown as was the case with the Temperature Module results. The humidity reading is near the lower middle of the list and shows a value of 16.5187. I am certain that the sensor is nowhere near that accurate; however, the owfs software calculates a value based on the raw data sent back to it from the module.

The last module is the Barometer Module with its results shown by clicking on the directory named 28.827BBC000000. Figure 10-21 shows the data on this page.

However, there is a problem here. If you look closely at all the data in this figure, you will not find any barometric pressure value. You must click on the directory entry named *B1-R1_A* to view another page with the desired data. Figure 10-22 shows this page.

The pressure displayed is 1001.87 millibars. The actual pressure measured at a nearby airport was 1014.6 millibars, so the Barometer Module was reading a bit low. The potentiometer mentioned in the module discussion needs to be adjusted to properly calibrate the barometric sensor.

Packet Sniffing

I connected the 1-Wire Sniffer Module to the network to test out the functionality. The sniffer was connected to a 64-bit laptop by using a USB-to-serial-port adapter. The laptop was running the Windows 7 Ultimate OS. I also used the Tera Term program that I previously discussed in Chap. 5 with a terminal configuration as listed in the above sniffer module discussion.

26.882728010000

uncached version

up	directory
B1-R1-A	B1-R1-A
CA	☑ CHANGE
EE	☑ CHANGE
HIH3600	HIH3600
HIH4000	HIH4000
HTM1735	HTM1735
IAD	☑ CHANGE
MultiSensor	MultiSensor
S3-R1-A	S3-R1-A
VAD	1.31
VDD	4.97
address	2688272801000044
alias	
crc8	44
date	Wed Dec 31 23:16:33 196 CHANGE
disconnect	disconnect
endcharge	endcharge
family	26
humidity	16.5187
id	882728010000
locator	FFFFFFFFFFFFFFFF
offset	1 CHANGE
pages	pages
r_address	4400000128278826
r_id	000001282788
r_locator	FFFFFFFFFFFFFFFF
temperature	24.375
type	DS2438
udate	15393 CHANGE
vis	0

Figure 10-20 Humidity Module results.

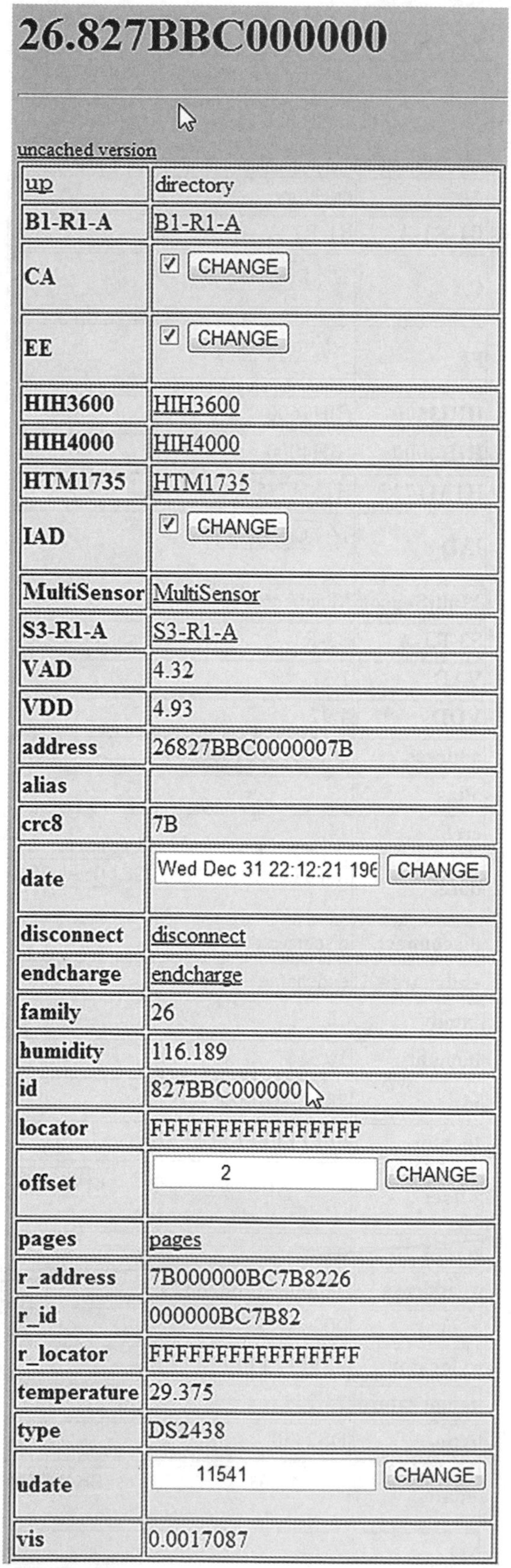

26.827BBC000000

uncached version

up	directory
B1-R1-A	B1-R1-A
CA	☑ CHANGE
EE	☑ CHANGE
HIH3600	HIH3600
HIH4000	HIH4000
HTM1735	HTM1735
IAD	☑ CHANGE
MultiSensor	MultiSensor
S3-R1-A	S3-R1-A
VAD	4.32
VDD	4.93
address	26827BBC0000007B
alias	
crc8	7B
date	Wed Dec 31 22:12:21 196 CHANGE
disconnect	disconnect
endcharge	endcharge
family	26
humidity	116.189
id	827BBC000000
locator	FFFFFFFFFFFFFFFF
offset	2 CHANGE
pages	pages
r_address	7B000000BC7B8226
r_id	000000BC7B82
r_locator	FFFFFFFFFFFFFFFF
temperature	29.375
type	DS2438
udate	11541 CHANGE
vis	0.0017087

Figure 10-21 Barometer Module results.

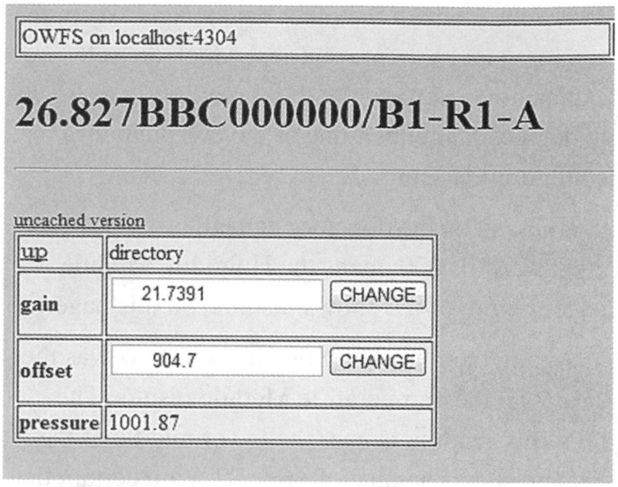

OWFS on localhost:4304

26.827BBC000000/B1-R1-A

uncached version

up	directory
gain	21.7391 CHANGE
offset	904.7 CHANGE
pressure	1001.87

Figure 10-22 Barometric pressure value.

I also decided to test the remote login capability that required me to determine my home network's public IP address. The procedure detailed in Chap. 8 was used. I also set the home network router's port forwarding to 2121 and pointed it to the RasPi's local address. Figure 10-23 is a combined screenshot that shows both the owfs opening page along with the Tera Term display of packet data that was created by this action.

The interpretation of the first line of data follows:

First data packet = RP F0 28 C1 EA 25 03 00 00 C4

RP is short for presence pulse.

F0 is a one byte ROM command. This one returned the ROM module ID.

28 C1 EA 25 03 00 00 is the ROM ID sent back to the master.

C4 is a CRC check byte used for error checking.

The following three lines in the sniffer window performed the same action for the other three modules in the 1-Wire network. Notice that they are displayed in precisely the same order as they are listed in the owfs directory web page.

I then clicked on the Temperature Module directory to see what the sniffer returned. Figure 10-24 is the resulting screenshot.

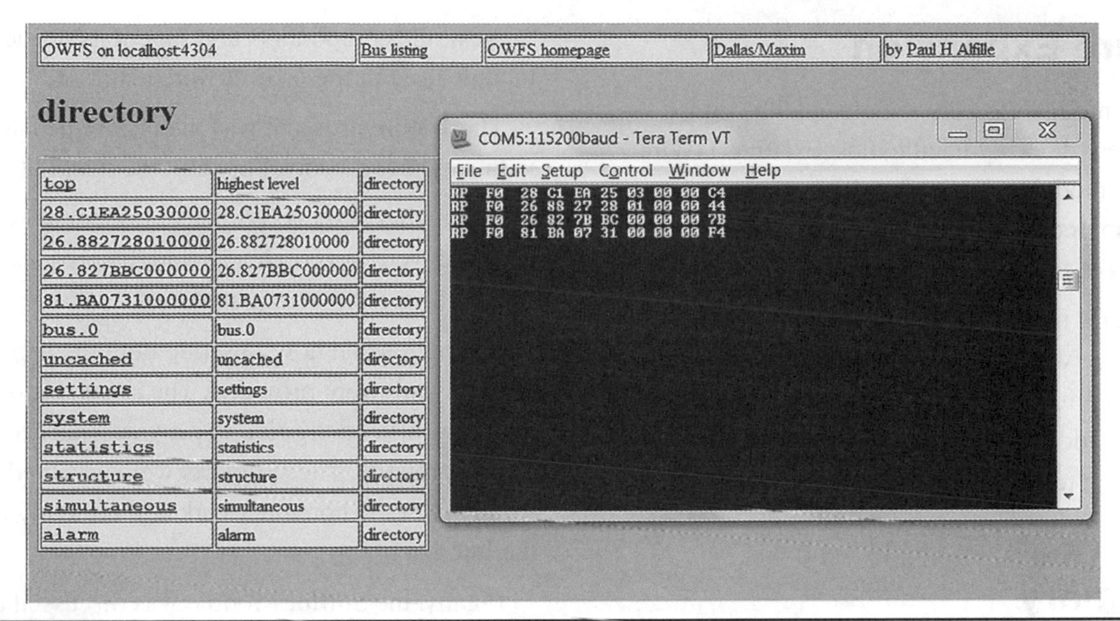

Figure 10-23 Screenshot of an owfs initial page overlaid with Tera Term packet data.

As one might expect, there is obviously a lot more information being created so that owfs can populate the Temperature Module's report page (Fig. 10-19). You should notice that the module's ROM ID is repeated in each line of the sniffer display. This makes it quite easy to identify the specific module that is transmitting data over the 1-Wire network. I refer you to the owfs user's manual if you desire to learn more about commands and returned data.

```
RP  F0  28 C1 EA 25 03 00 00 C4
RP  55  28 C1 EA 25 03 00 00 C4  BE  40  01  4B  46  7F  FF  10  10
 1D
RP  55  28 C1 EA 25 03 00 00 C4  4E  4B  46  1F
RP  55  28 C1 EA 25 03 00 00 C4  B4  FF
RP  55  28 C1 EA 25 03 00 00 C4  44  00  00  FF
RP  55  28 C1 EA 25 03 00 00 C4  BE  40  01  4B  46  1F  FF  10  10
 8D
RP  F0  28 C1 EA 25 03 00 00 C4 FC FF FF
RP  F0  28 C1 EA 25 03 00 00 C4 FC FF FF
RP  55  28 C1 EA 25 03 00 00 C4  BE  40  01  4B  46  1F  FF  10  10
 8D
RP  55  28 C1 EA 25 03 00 00 C4  4E  4B  46  7F
RP  55  28 C1 EA 25 03 00 00 C4  44  00  00  00  00  00  00  00  00
 00  00  00  00  FF
RP  55  28 C1 EA 25 03 00 00 C4  BE  3E  01  4B  46  7F  FF  02  10
 6C
RP  55  28 C1 EA 25 03 00 00 C4  BE  3E  01  4B  46  7F  FF  02  10
 6C
RP  55  28 C1 EA 25 03 00 00 C4  4E  4B  46  3F
RP  55  28 C1 EA 25 03 00 00 C4  44  00  00  00  FF
RP  55  28 C1 EA 25 03 00 00 C4  BE  40  01  4B  46  3F  FF  10  10
 FD
RP  55  28 C1 EA 25 03 00 00 C4  BE  40  01  4B  46  3F  FF  10  10
 FD
RP  55  28 C1 EA 25 03 00 00 C4  4E  4B  46  5F
RP  55  28 C1 EA 25 03 00 00 C4  44  00  00  00  00  00  00  FF
RP  55  28 C1 EA 25 03 00 00 C4  BE  3E  01  4B  46  5F  FF  02  10
 1C
RP  55  28 C1 EA 25 03 00 00 C4  B8
RP  55  28 C1 EA 25 03 00 00 C4  BE  3E  01  4B  46  7F  FF  02  10
 6C
RP  55  28 C1 EA 25 03 00 00 C4  B8
RP  55  28 C1 EA 25 03 00 00 C4  BE  3E  01  4B  46  7F  FF  02  10
 6C
```

Figure 10-24 Screenshot after the Temperature Module directory was clicked.

Future Expansion

Adding modules is very simple, and the owfs software will easily handle most existing 1-Wire weather modules. You are really limited by your own resources, if you desire to add additional modules to the 1-Wire network.

A useful module to add would be an anemometer to measure wind speed and a compass encoder to get wind direction. The only other module that could be added to fully populate this weather station is a rain gauge.

Summary

The chapter began with a review of the weather station design that is based upon using a series of separate, interconnected sensor modules, all using the 1-Wire network.

Next, the components and functions of each module used in the project were discussed.

The 1-Wire protocol was then examined along with an integral controller chip, the DS2438A.

Various ways the 1-Wire network could be powered were also reviewed and the pros and cons of each approach were pointed out.

Instructions on how to set up both the hardware and software were provided. The software package owfs was also configured to provide Web-based weather reports. These reports were accessed both locally over the home network and remotely over the Internet.

Finally, the Sniffer Module was discussed as a means of checking real-time data flowing over the 1-Wire network. This is a very handy tool to have available for debugging and for program development.

CHAPTER 11

Local and Remote Logic Controller

Introduction

This project consists of two major parts both of which concern logic control, or put simply, turning IO pins on and off. The two project parts are closely related, and both involve the control of IO pins: one uses an external board that provides eight additional GPIO pins, and the other is based on the RasPi itself.

The external board uses the 1-Wire protocol that was introduced in the Chap. 10 project for building a weather station. This board may be either controlled through a program running on the RasPi or accessed using the home network, as was demonstrated in the Chap. 10 weather project.

The other project part directly controls the RasPi GPIO pins. This part may be thought of as an extension of the fundamental concepts that were first discussed in the Chap. 2 demonstration of a blinking LED. In this part, the GPIO pins will be controlled from an Android smartphone client application, with the RasPi acting as a web server. The RasPi may also be controlled from a browser, which will also be demonstrated.

At the end of the chapter, I will discuss how both project parts can be accessed over the Internet .

The good news regarding this project is that there is very little hardware construction involved. The more interesting (never bad) news is that a fair amount of software installation and configuration is involved; however, at this stage in the book, you should be fairly comfortable dealing with such activities. Now we'll look at the 1-Wire external control board.

1-Wire External Control Board

The 1-Wire board I'm using is the Hobby Boards 8 Channel I/O v2.0 board, as shown in Fig. 11-1. It is part of the series "29" family and is fully compliant with the 1-Wire protocol.

Significant board specifications are listed in Table 11-1.

An eight-position DIP switch labeled "Relay Control" is on the board. It can disable any of the relays if you wish to do so. Another eight-position DIP switch labeled "Input Ground" is used in connection with the input terminals. Inputs should be connected as follows:

1. **Non-powered inputs**—Connect one lead to the positive terminal marked "+." The other lead is connected to the powered screw terminal marked "+5v." The corresponding "Input Ground" DIP switch should be in the on (up) position, which is also the default position.
2. **Powered inputs**—Connect one lead to the positive terminal marked "+." The other lead is connected to the negative terminal marked "–." The corresponding "Input Ground" DIP switch should be in the off (down) position. The input polarity is a "don't care" situation because the input opto-isolators are bidirectional, as can

Figure 11-1 Hobby Boards 8 Channel I/O v2.0 board.

be seen in Fig. 11-2, which shows a portion of the board's schematic. You can even use low-voltage AC to trigger the input.

The full schematic along with the board's user's manual is available from www.hobby-boards.com.

Each channel relay is connected in parallel with the 1-Wire controller and the respective input for that channel. This means that the relay can be activated by a network command or by an input action, such as a contact closure. I will demonstrate this dual action in a later section of this chapter.

A block diagram of the 1-Wire network used in this project is shown in Fig. 11-3. The USB to

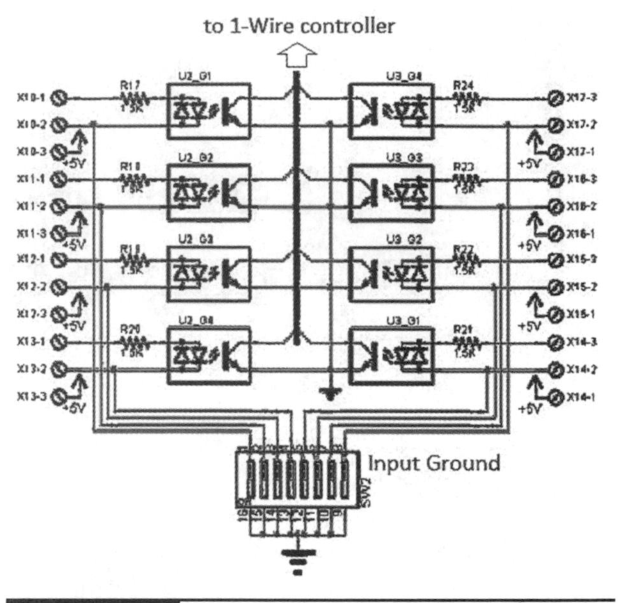

Figure 11-2 8 Channel I/O board input schematic.

1-Wire adapter and 1-Wire Sniffer Module used in the Chap. 10 project are also being used in this project. The 8 Channel I/O board is also configured to supply power to the entire 1-Wire network.

All the modules are interconnected using Ethernet patch cables except for the special RJ12/RJ45 cable that connects the USB/1-Wire adapter and the I/O board.

Table 11-1 8 Channel I/O Board Specifications

Component	Description
Controller	Dallas Semiconductor DS2408 1-Wire chip
Connections	Dual RJ45 connectors for the 1-Wire network All I/O connections to the board use screw terminals 2.1 mm barrel type power connector 5 V DC screw connections available for external use (100 mA max)
Outputs	8 SPDT relays with LED indicators. Max rating .25 A @ 250 V AC Each relay has Common (c), Normally Open (NO) and Normally Closed (NC) terminals connected to its own screw terminal
Inputs	8 opto-isolated inputs Each input has the positive (+), negative (–) and +5 V DC (+5v) terminal connected to its own screw terminal
Power Required	250 ma @ 15 V DC. Provided by 15 V DC wall wart supply jumper option to power the 1-Wire network

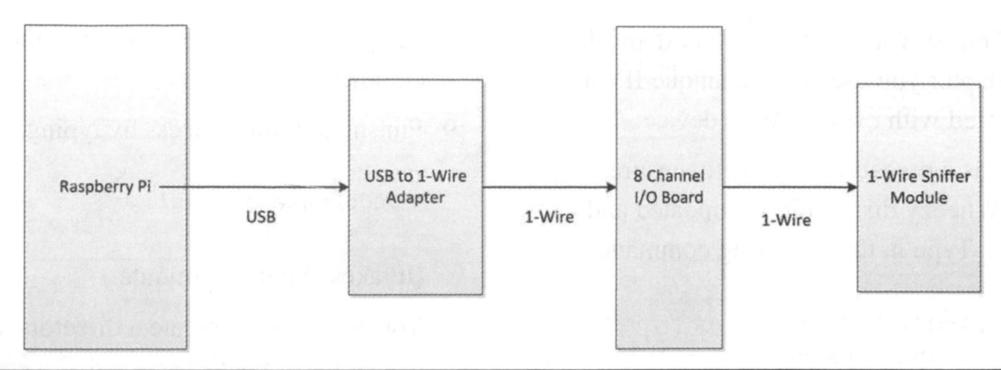

Figure 11-3 Block diagram of the 1-Wire network.

1-Wire File System (owfs) Installation and Configuration

The software required to control the 8 Channel I/O board needs to be built from the owfs source code in a procedure similar to that followed in Chap. 9 for the libnfc software. This time, I will not include the many screenshots shown in Chap. 9, since you should now be somewhat familiar with the build process. I do caution you to carefully follow all the steps because it is quite easy to miss something, and consequently, to be unsuccessful in creating the operational software. The prerequisite conditions for the build are to have an Internet-connected RasPi running the latest Wheezy distribution with a powered USB hub attached. You can also choose to login remotely by using an SSH connection, but that is not a requirement.

1. The first step is to simply plug the USB/ 1-Wire adapter into the powered hub and confirm that the Wheezy OS recognizes the adapter. You should also attach the remaining modules at this time. Then type the following at the command line prompt:

```
lsusb ↵
```

Figure 11-4 shows what displayed on my setup. The USB/1-Wire adapter is shown as "Bus 001 Device 005: ID 04fa:2490 Dallas Semiconductor DS1490F 2-in-1 Fob, 1-Wire

```
pi@raspberrypi: ~
login as: pi
pi@192.168.1.21's password:
Linux raspberrypi 3.6.11+ #371 PREEMPT Thu Feb 7 16:31:35 GMT 2013 armv6l

The programs included with the Debian GNU/Linux system are free software;
the exact distribution terms for each program are described in the
individual files in /usr/share/doc/*/copyright.

Debian GNU/Linux comes with ABSOLUTELY NO WARRANTY, to the extent
permitted by applicable law.
Last login: Fri Mar  8 19:22:13 2013
pi@raspberrypi ~ $ ls
Desktop  ocr_pi.png  python_games
pi@raspberrypi ~ $ lsusb
Bus 001 Device 002: ID 0424:9512 Standard Microsystems Corp.
Bus 001 Device 001: ID 1d6b:0002 Linux Foundation 2.0 root hub
Bus 001 Device 003: ID 0424:ec00 Standard Microsystems Corp.
Bus 001 Device 004: ID 05e3:0608 Genesys Logic, Inc. USB-2.0 4-Port HUB
Bus 001 Device 005: ID 04fa:2490 Dallas Semiconductor DS1490F 2-in-1 Fob, 1-Wire
 adapter
Bus 001 Device 006: ID 1241:1177 Belkin F8E842-DL Mouse
Bus 001 Device 007: ID 046d:c326 Logitech, Inc.
pi@raspberrypi ~ $
```

Figure 11-4 lsusb display.

adapter" entry. Your entry will vary depending on the adapter you use and the unique ID that is associated with every 1-Wire device.

2. This step is a precautionary one to ensure that the Wheezy distribution is updated and upgraded. Type in the following commands:

```
sudo apt-get update ↵
sudo apt-get upgrade ↵
```

(This can take some time if there are many upgrades needed.)

3. Next, the required build tools need to be installed. Type in the following command:

```
sudo apt-get install autoconf
libtool libusb-dev libfuse-dev ↵
```

4. The source code has to be downloaded next. Type in the following command:

```
sudo wget http://sourceforge.net/
projects/owfs/files/owfs/2.8p15/
owfs-2.8p15.tar.gz ↵
```

5. Extract the source code and subdirectory from the archive file by typing:

```
sudo tar zxvf owfs-2.8p15.tar.gz ↵
```

6. Change from the "pi" directory to the newly created "owfs-2.8p15" directory by typing:

```
cd owfs-2.8p15 ↵
```

7. The source code must now be "configured." Type:

```
sudo ./configure ↵
```

(This step takes about three minutes to complete.)

8. The build happens next. Type:

```
sudo make ↵
```

(Be patient; this takes over 30 minutes to complete.)

9. Finish the build process by typing:

```
sudo make install ↵
```

(It takes about one minute.)

10. You now need to create a directory where the device files may be accessed. Type:

```
sudo mkdir /mnt/1wire ↵
```

11. I found the next step useful as a way to ensure that the RasPi was in a good or consistent state. Type:

```
sudo reboot ↵
```

Login as you would normally, and you should be in the "pi" directory.

12. You now need to start the owfs filesystem and identify the mount point. Type the following:

```
sudo /opt/owfs/bin/owfs —allow_
other —u —m /mnt/1wire ↵
```

NOTE This command should be done only once after the initial boot. If you try to remount the 1wire directory from the command line prompt after an initial mount operation, it typically results in a "Permission Denied" error. However, I did find that remounts were possible if used in a program statement. So-called "nonempty" warnings were created by the owfs file system.

Figure 11-5 is a screenshot showing the results of the mount operation. I then changed directories from "pi" to "1wire" by typing:

```
cd /mnt/1wire ↵
```

I did an "`ls`" to display the 1wire directory contents that identified the two normal 1-Wire

```
pi@raspberrypi ~/owfs-2.8p15 $ sudo mkdir /mnt/1wire
pi@raspberrypi ~/owfs-2.8p15 $ sudo /opt/owfs/bin/owfs --allow_other -u -m /mnt/1wi
re
DEFAULT: ow_usb_msg.c:(295) Opened USB DS9490 bus master at 1:5.
DEFAULT: ow_usb_cycle.c:(191) Set DS9490 1:5 unique id to 81 BA 07 31 00 00 00 F4
pi@raspberrypi ~/owfs-2.8p15 $ cd /mnt/1wire
pi@raspberrypi /mnt/1wire $ ls
29.B82E08000000  alarm  settings    structure  uncached
81.BA0731000000  bus.0  statistics  system
pi@raspberrypi /mnt/1wire $ cd 29.B82E08000000
pi@raspberrypi /mnt/1wire/29.B82E08000000 $ ls
address  latch.1  latch.7     PIO.0  PIO.6     r_address  sensed.3    sensed.BYTE
alias    latch.2  latch.ALL   PIO.1  PIO.7     r_id       sensed.4    set_alarm
crc8     latch.3  latch.BYTE  PIO.2  PIO.ALL   r_locator  sensed.5    strobe
family   latch.4  LCD_H       PIO.3  PIO.BYTE  sensed.0   sensed.6    type
id       latch.5  LCD_M       PIO.4  por       sensed.1   sensed.7
latch.0  latch.6  locator     PIO.5  power     sensed.2   sensed.ALL
pi@raspberrypi /mnt/1wire/29.B82E08000000 $
```

Figure 11-5 1-Wire owfs mount operation.

devices on the network (the Sniffer Module is non-conforming). The USB/1-Wire adapter belongs to the family of 1-Wire devices beginning with the string "81." That meant the remaining directory, starting with the family designator of "29" must be the 8 Channel I/O board.

I changed into that directory by typing:

```
cd 29.B82E08000000 ↵
```

The directory contents are displayed at the bottom of Fig. 11-4.

I would like next to clarify an important aspect of how Linux treats hardware before procceding much further into this part of the project.

Linux, Hardware, and FUSE

Unix and its descendent Linux "treat everything as files" according to the old mantra. This phrase encompasses all things, including hard drive files, serial ports, thumb drives, etc. You can read and write to a hardware device, if it accommodates such actions, in exactly the same way as you can read and write to an ordinary file, provided the permissions are properly set. The view that everything can be handled as a file certainly simplifies how programs handle data to and from hardware devices.

There is also another particular file category that doesn't appear until the OS is running. In this category are virtual files that are dynamically created upon demand and are often-times stored in the /sys directory. The *Filesystem in Userspace* (FUSE) application supports the owfs application by creating and storing these virtual files in FUSE registered directories. The "/mnt/1wire directory" is the designated site in which to store all the virtual files created for the 1-Wire network currently operating.

You can prove to yourself that the virtual files exist. Simply reboot and do a "`ls`" on the /mnt/1wire directory. Nothing will show. Next type in the mount command from step 12 and redo the "`ls`" command. Then you should see a directory data display like the one shown in Fig. 11-4 for the 1wire directory. owfs applications and FUSE are integral to each other; the system wouldn't work without either of them.

Another very useful feature is that the 1wire directory will automatically be updated as modules are removed from and/or added to the 1-Wire network. Figure 11-6 shows what happened before and after I plugged a Temperature Module (family "28") into the 1-Wire network. Note, that I did not have to type any commands; the addition was automatically detected.

You will see that a new directory entry named "simultaneous" appeared in the listing. This new

```
pi@raspberrypi ~ $ cd /mnt/1wire
pi@raspberrypi /mnt/1wire $ ls
29.B82E08000000  alarm  settings    structure  uncached
81.BA0731000000  bus.0  statistics  system
pi@raspberrypi /mnt/1wire $ ls
28.C1EA25030000  81.BA0731000000  bus.0     simultaneous  structure  uncached
29.B82E08000000  alarm            settings  statistics    system
pi@raspberrypi /mnt/1wire $ 
```

Figure 11-6 Autodetection of 1-Wire Modules.

entry was created by owfs and FUSE after the programs recognized that the new module was part of the Temperature Module virtual file properties. This dynamic recognition and autoconfiguration is a very powerful concept, which contributes to making 1-Wire networks so easy to install and use.

Test Sequences for the 8 Channel I/O Board

What follows are some quick and simple terminal commands to test the basic board functions. Figure 11-7 shows the test setup, using a laptop that is controlling the RasPi with an SSH session, i.e., "running the Pi headless."

If you haven't yet done it, mount the 1wire directory, using the command from step 12 shown above. Remember: attempting to remount after doing it once will result in a Permission Denied error. To clear the error, all you have to do is reboot the RasPi and reissue the mount command.

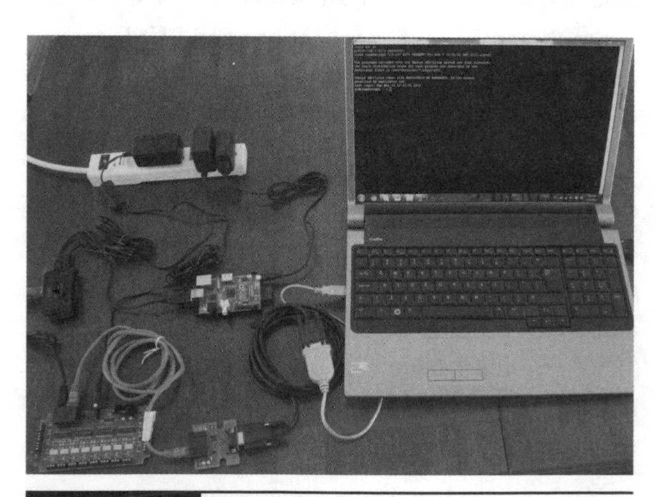

Figure 11-7 1-Wire test setup.

Next, you need to change directories to be in the 8 Channel IO board directory. For my system the command entered was:

```
cd /mnt/1wire/29.B82E08000000 ↵
```

Your command will be slightly different because the board's unique serial number and family name make up the directory name. Simply "`cd`" into /mnt/1wire and then "`ls`" if you haven't previously recorded the board's directory name.

Testing the Outputs

Once you are safely in the device directory, do another "`ls`" to see all the different files owfs and FUSE have created. Then type the following:

```
echo '255' > PIO.BYTE ↵
```

You should be rewarded by hearing all eight relays click and seeing all eight LEDs light up. This command simply sends a byte of all "1's" into the file named PIO.BYTE that controls all eight relays simultaneously. Tired of seeing all the LEDs lit? Type the following:

```
echo '0' > PIO.BYTE ↵
```

All the LEDs should be off now and all the relays deactivated.

Of course, specific relays may be activated by sending the appropriate bit pattern encoded into the equivalent decimal number to the PIO.BYTE file. For instance, if you wanted to turn on relays 1, 3, 5, and 7, you would need to send 85 to PIO.BYTE. The number 85 was calculated by using the binary weights listed in Table 11-2.

To turn on relays 1, 3, 5, and 7, all that is needed is to sum the binary weights associated with each relay:

Relay Number	Weight
1	1
3	4
5	16
7	64
Sum	85

Type the command:

```
echo '85' > PIO.BYTE ↵
```

Relays 1, 3, 5, and 7 should now be activated and their LEDs on. You now know how to turn them off.

Individual relays may also be controlled by using specific files dedicated to each relay. For example, to turn on relay 4, type the following:

```
echo '1' > PIO.3 ↵
```

Relay 4 turns on. The individual relay file names start at 0 not 1. This is not true for the board relays, which are numbered 1 through 8. This difference is a small detail that can nonetheless cause some confusion.

Table 11-2 Relay Binary Weights

Relay #	1	2	3	4	5	6	7	8
Binary weight	1	2	4	8	16	32	64	128

Testing the Inputs

Sensing inputs is very similar to setting outputs except for one major difference. A '1' is used to indicate that no input is present or detected. All the relays must also be disabled, by turning all the switches on the Relay Control DIP switch to the off (down) position. This configuration is required because an IO line cannot be both an input and an output at the same time.

Next, ensure that nothing is connected to any input and then type the following:

```
cat sensed.BYTE ↵
```

You should see the value 255 displayed just before the command line prompt, just like the circled number in Fig. 11-8.

The value 255 is the decimal equivalent of all 1's for 8 bits or 1 byte. To test this concept out a bit further, connect a jumper wire between the screw terminals marked "+5v" and "+" on input terminal 8. What would you expect to see displayed after you type in "cat sensed.BYTE"? A moment's thought might help you realize that the binary value of the bit for the number 8 input (relay binary weight of 128) would be 0; hence the value should be 255–128 or 127. Go ahead and type the command:

```
cat sensed.BYTE ↵
```

You should be rewarded to see the value 127, just as you expected. It's not really too hard after all!

Now it is time to try out a Python program to exercise the 8-Channel I/O board. You first need

Figure 11-8 Result for the sensed.BYTE command.

to enable all the relays by putting all the Relay Control DIP switches to the on (up) position.

Python Test Program

This program cycles relays 2 and 4 every 10 seconds forever unless stopped by a keyboard interrupt that is done by pressing the 'control' and 'c' keys simultaneously (^C). I also included a statement to autoload the 1wire directory so that the command shown in step 12 does not have to be explicitly entered. The file object statement will have to be modified to suit your installation, since it has the unique address for my 8-Channel I/O Board hard coded into it. This file is named *ow_eight.py* and is available from the book's companion website, www.mhprofessional.com/raspi.

Type the following at the command line prompt:

```
sudo python ow_eight.py ↵
```

Relays 2 and 4 should click, and the associated LEDs should blink on for 5 seconds then off for 5 seconds. This program should be used as a starting point from which you can develop your own control program to match your desired application.

Remember that a given I/O channel may be used either as an input or an output but not as both simultaneously. This is precisely the situation with the RasPi GPIO pins where a particular pin may be set as an output for one part of an application and then used as an input in another part. The major difference between the RasPi and the I/O board is that a physical switch must be set on the I/O board, while a GPIO pin may be reset by a program statement with the RasPi. It is just a bit more limiting to use the I/O board as compared to the pure programming situation allowed with the RasPi. However, having eight channels available provides a good deal of flexibility.

Caution **The relay contacts are rated for a maximum of 250 mA at 250 V AC, which equates to an approximate 60-watt load. Do not attempt to switch more than a 60-watt load using these light duty relays, and under *NO* circumstances should you try to control a line voltage AC motor. If you want to control a heavy AC-inductive load, use the relay to control a motor controller that is rated for the AC load. This is a safe practice, and one that should be adhered to at all times.**

ow_eight.py

```
import time
import os
ow_id = os.system('sudo /opt/owfs/bin/owfs --allow_other -u -m /mnt/1wire')
#file object statement—modify the device directory to match your own board ID.
file = os.path.join('/','mnt', '1wire', '29.B82E08000000', 'PIO.BYTE')
while True:
    file_obj = open(file, 'r+')
    file_obj.write('10')
    line = file_obj.read()
    print(line)
    time.sleep(5)
    file_obj.write('0')
    line = file_obj.read()
    time.sleep(5)
    file_obj.close()
```

Sniffer Monitoring

I was interested in monitoring the 1-Wire network activity while the ow_eight.py program was running. Figure 11-9 is a screenshot from my laptop running the Tera Term program that I discussed in Chap. 10. The laptop was connected to the Sniffer Module, using a USB-to-serial adapter and cable.

The figure shows the repeated activity of switching the selected channels on and off, which was expected. What was not expected is the periodic polling of the 1-Wire network taking place every two minutes. That may be deduced by observing the shorter packets interspersed among the repeated control packets initiated by the ow_eight program. The two-minute interval was calculated by knowing that each repeat packet was issued every 10 seconds.

While not critical for program development, the sniffer is a very handy tool that helps you understand what is occurring within the 1-Wire network.

Android Remote Control

In this section I will show you how to control the RasPi GPIO pins by using an application running on an Android smartphone. The smartphone application is named *DrGPIO* and is available as *donationware* from the Google's Play Store app. Donationware, as the name implies, means that you may donate to the author if you use it and find it useful. You must also download the web server portion named WebIOPi. Follow this procedure to download and install the server:

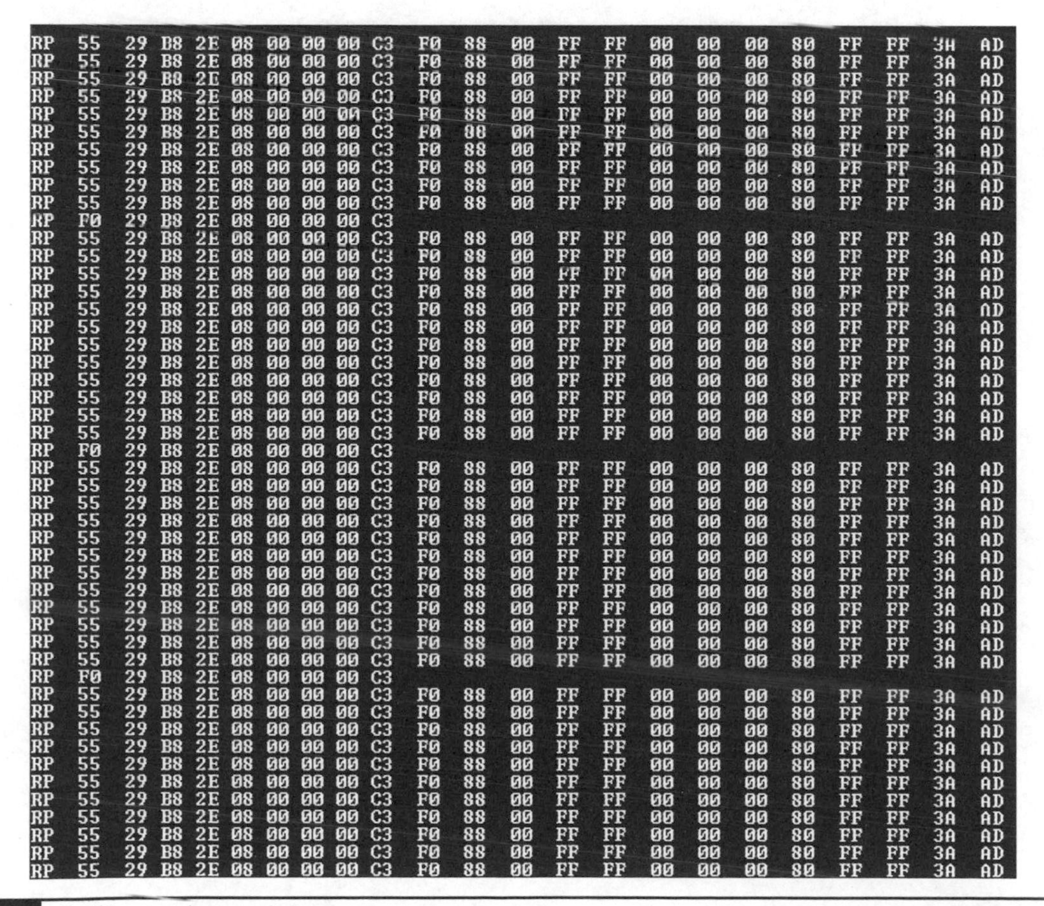

```
RP 55 29 D8 2E 08 00 00 00 C3 F0 88 00 FF FF 00 00 00 80 FF FF 3H AD
RP 55 29 B8 2E 08 00 00 00 C3 F0 88 00 FF FF 00 00 00 80 FF FF 3A AD
RP 55 29 B8 2E 08 00 00 00 C3 F0 88 00 FF FF 00 00 00 80 FF FF 3A AD
RP 55 29 B8 2E 08 00 00 00 C3 F0 88 00 FF FF 00 00 00 80 FF FF 3A AD
RP 55 29 B8 2E 08 00 00 00 C3 F0 88 00 FF FF 00 00 00 80 FF FF 3A AD
RP 55 29 B8 2E 08 00 00 00 C3 F0 88 00 FF FF 00 00 00 80 FF FF 3A AD
RP 55 29 B8 2E 08 00 00 00 C3 F0 88 00 FF FF 00 00 00 80 FF FF 3A AD
RP 55 29 B8 2E 08 00 00 00 C3 F0 88 00 FF FF 00 00 00 80 FF FF 3A AD
RP 55 29 B8 2E 08 00 00 00 C3 F0 88 00 FF FF 00 00 00 80 FF FF 3A AD
RP 55 29 B8 2E 08 00 00 00 C3 F0 88 00 FF FF 00 00 00 80 FF FF 3A AD
RP F0 29 B8 2E 08 00 00 00 C3
RP 55 29 B8 2E 08 00 00 00 C3 F0 88 00 FF FF 00 00 00 80 FF FF 3A AD
RP 55 29 B8 2E 08 00 00 00 C3 F0 88 00 FF FF 00 00 00 80 FF FF 3A AD
RP 55 29 B8 2E 08 00 00 00 C3 F0 88 00 FF FF 00 00 00 80 FF FF 3A AD
RP 55 29 B8 2E 08 00 00 00 C3 F0 88 00 FF FF 00 00 00 00 FF FF 3A AD
RP 55 29 B8 2E 08 00 00 00 C3 F0 88 00 FF FF 00 00 00 80 FF FF 3A AD
RP 55 29 B8 2E 08 00 00 00 C3 F0 88 00 FF FF 00 00 00 80 FF FF 3A AD
RP 55 29 B8 2E 08 00 00 00 C3 F0 88 00 FF FF 00 00 00 80 FF FF 3A AD
RP 55 29 B8 2E 08 00 00 00 C3 F0 88 00 FF FF 00 00 00 80 FF FF 3A AD
RP 55 29 B8 2E 08 00 00 00 C3 F0 88 00 FF FF 00 00 00 80 FF FF 3A AD
RP 55 29 B8 2E 08 00 00 00 C3 F0 88 00 FF FF 00 00 00 80 FF FF 3A AD
RP 55 29 B8 2E 08 00 00 00 C3 F0 88 00 FF FF 00 00 00 80 FF FF 3A AD
RP 55 29 B8 2E 08 00 00 00 C3 F0 88 00 FF FF 00 00 00 80 FF FF 3A AD
RP F0 29 B8 2E 08 00 00 00 C3
RP 55 29 B8 2E 08 00 00 00 C3 F0 88 00 FF FF 00 00 00 80 FF FF 3A AD
RP 55 29 B8 2E 08 00 00 00 C3 F0 88 00 FF FF 00 00 00 80 FF FF 3A AD
RP 55 29 B8 2E 08 00 00 00 C3 F0 88 00 FF FF 00 00 00 80 FF FF 3A AD
RP 55 29 B8 2E 08 00 00 00 C3 F0 88 00 FF FF 00 00 00 80 FF FF 3A AD
RP 55 29 B8 2E 08 00 00 00 C3 F0 88 00 FF FF 00 00 00 80 FF FF 3A AD
RP 55 29 B8 2E 08 00 00 00 C3 F0 88 00 FF FF 00 00 00 80 FF FF 3A AD
RP 55 29 B8 2E 08 00 00 00 C3 F0 88 00 FF FF 00 00 00 80 FF FF 3A AD
RP 55 29 B8 2E 08 00 00 00 C3 F0 88 00 FF FF 00 00 00 80 FF FF 3A AD
RP 55 29 B8 2E 08 00 00 00 C3 F0 88 00 FF FF 00 00 00 80 FF FF 3A AD
RP 55 29 B8 2E 08 00 00 00 C3 F0 88 00 FF FF 00 00 00 80 FF FF 3A AD
RP 55 29 B8 2E 08 00 00 00 C3 F0 88 00 FF FF 00 00 00 80 FF FF 3A AD
RP 55 29 B8 2E 08 00 00 00 C3 F0 88 00 FF FF 00 00 00 80 FF FF 3A AD
RP F0 29 B8 2E 08 00 00 00 C3
RP 55 29 B8 2E 08 00 00 00 C3 F0 88 00 FF FF 00 00 00 80 FF FF 3A AD
RP 55 29 B8 2E 08 00 00 00 C3 F0 88 00 FF FF 00 00 00 80 FF FF 3A AD
RP 55 29 B8 2E 08 00 00 00 C3 F0 88 00 FF FF 00 00 00 80 FF FF 3A AD
RP 55 29 B8 2E 08 00 00 00 C3 F0 88 00 FF FF 00 00 00 80 FF FF 3A AD
RP 55 29 B8 2E 08 00 00 00 C3 F0 88 00 FF FF 00 00 00 80 FF FF 3A AD
RP 55 29 B8 2E 08 00 00 00 C3 F0 88 00 FF FF 00 00 00 80 FF FF 3A AD
RP 55 29 B8 2E 08 00 00 00 C3 F0 88 00 FF FF 00 00 00 80 FF FF 3A AD
RP 55 29 B8 2E 08 00 00 00 C3 F0 88 00 FF FF 00 00 00 80 FF FF 3A AD
RP 55 29 B8 2E 08 00 00 00 C3 F0 88 00 FF FF 00 00 00 80 FF FF 3A AD
RP 55 29 B8 2E 08 00 00 00 C3 F0 88 00 FF FF 00 00 00 80 FF FF 3A AD
RP 55 29 B8 2E 08 00 00 00 C3 F0 88 00 FF FF 00 00 00 80 FF FF 3A AD
```

Figure 11-9 Sniffer packet activity.

1. Type the following at a command line prompt:

```
sudo wget http://webiopi
.googlecode.com/files/WebIOPi
-0.5.3.tar.gz ↵
```

2. Next, extract the source code:

```
sudo tar xvzf WebIOPi-0.5.3.tar
.gz ↵
```

3. Change directories to the newly created one:

```
cd WebIOPi-0.5.3 ↵
```

4. Install and set up the software using a preset script:

```
sudo ./setup.sh ↵
```

I had the initial setup fail on me. I did the following to resolve the issue:

```
sudo apt-get update ↵
sudo apt-get python3.2-dev ↵
```

Repeat step 4 above.

Figure 11-10 shows the end result of the setup command.

5. The HTTP web server is started using Python as follows:

```
sudo python –m webiopi 8000 ↵
```

The 8000 is optional because it is the default. I like to include it, since it is a reminder of the

```
pi@raspberrypi: ~/WebIOPi-0.5.3
File Edit Tabs Help
byte-compiling build/bdist.linux-armv6l/egg/_webiopi/GPIO.py to GPIO.cpython-32.
pyc
byte-compiling build/bdist.linux-armv6l/egg/_webiopi/__init__.py to __init__.cpy
thon-32.pyc
creating build/bdist.linux-armv6l/egg/EGG-INFO
copying WebIOPi.egg-info/PKG-INFO -> build/bdist.linux-armv6l/egg/EGG-INFO
copying WebIOPi.egg-info/SOURCES.txt -> build/bdist.linux-armv6l/egg/EGG-INFO
copying WebIOPi.egg-info/dependency_links.txt -> build/bdist.linux-armv6l/egg/EG
G-INFO
copying WebIOPi.egg-info/top_level.txt -> build/bdist.linux-armv6l/egg/EGG-INFO
writing build/bdist.linux-armv6l/egg/EGG-INFO/native_libs.txt
zip_safe flag not set; analyzing archive contents...
_webiopi.__pycache__.GPIO.cpython-32: module references __file__
creating 'dist/WebIOPi-0.5.3-py3.2-linux-armv6l.egg' and adding 'build/bdist.lin
ux-armv6l/egg' to it
removing 'build/bdist.linux-armv6l/egg' (and everything under it)
Processing WebIOPi-0.5.3-py3.2-linux-armv6l.egg
removing '/usr/local/lib/python3.2/dist-packages/WebIOPi-0.5.3-py3.2-linux-armv6
l.egg' (and everything under it)
creating /usr/local/lib/python3.2/dist-packages/WebIOPi-0.5.3-py3.2-linux-armv6l
.egg
Extracting WebIOPi-0.5.3-py3.2-linux-armv6l.egg to /usr/local/lib/python3.2/dist
-packages
WebIOPi 0.5.3 is already the active version in easy-install.pth

Installed /usr/local/lib/python3.2/dist-packages/WebIOPi-0.5.3-py3.2-linux-armv6
l.egg
Processing dependencies for WebIOPi==0.5.3
Finished processing dependencies for WebIOPi==0.5.3
WebIOPi installed for Python 3.2.3

Copying resources...

WebIOPi successfully installed
* To start WebIOPi with python  : sudo python -m webiopi
* To start WebIOPi with python3 : sudo python3 -m webiopi

* To start WebIOPi at boot      : sudo update-rc.d webiopi defaults
* To start WebIOPi service      : sudo /etc/init.d/webiopi start

* Look in /home/pi/WebIOPi-0.5.3/examples for Python library usage examples

pi@raspberrypi ~/WebIOPi-0.5.3 $
```

Figure 11-10 Setup command end result.

```
pi@raspberrypi ~/WebIOPi-0.5.3 $ sudo python -m webiopi 8000
WebIOPi/Python2/0.5.3 Access protected using /etc/webiopi/passwd
WebIOPi/Python2/0.5.3 HTTP Server started at http://192.168.1.21:8000/webiopi/
```

Figure 11-11 Starting the WebIOPi HTTP web server.

actual port number. Figure 11-11 shows the console display for the web server starting.

The user name is "webiopi" and password is "raspberry" if you are asked for them by the application. The service may be stopped by either a keyboard interrupt (^C) or closing the terminal window in which it was started. You may also start and stop the web server as a service using these commands:

```
sudo /etc/init.d/webiopi start ↵
sudo /etc/init.d/webiopi stop ↵
```

The web server can even be started during the boot process by entering the following command:

```
sudo update-rc.d webiopi defaults ↵
```

Testing the Web server with the Android App

I set up the RasPi with a Pi Cobbler prototype tool to drive three LEDs, as shown in the Fig. 11-12 diagram.

You will need to determine the RasPi's local IP address in order to connect the Android application to the RasPi web server. In my case, it was 192.168.1.21. All I needed to do was type that address into the phone, as shown in Fig. 11-13.

Scroll down to the bottom of the phone screen after you have entered the needed information. Tap on the button "Save and continue" and a pop-up dialog box will appear stating "DRGPIO is going to try to connect to your Raspberry Pi now." Tap the OK button, and you should now be shown a screen similar to the one in Fig. 11-14.

Figure 11-14 is a graphical representation of the RasPi 26-pin GPIO connector. The check boxes shown immediately to the side of each pin box allow you to select whether the pin is to act as an input or output. The check mark will be highlighted if it is set as an input. We need pins 18, 23, and 25 unchecked so that they are functioning as outputs. To output a 1, simply tap the pin box. The bar in

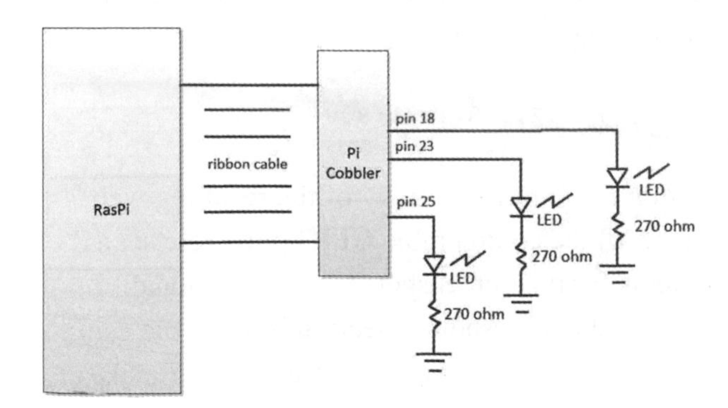

Figure 11-12 Test setup diagram.

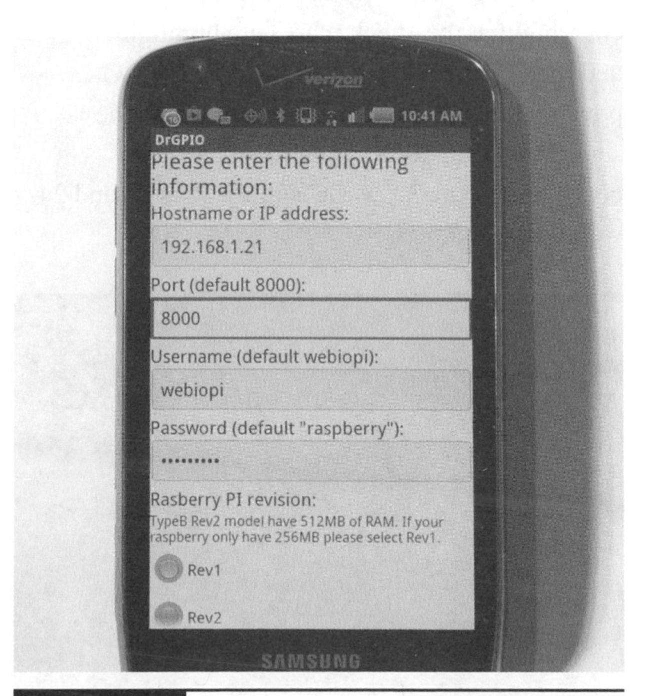

Figure 11-13 Android smartphone setup screen.

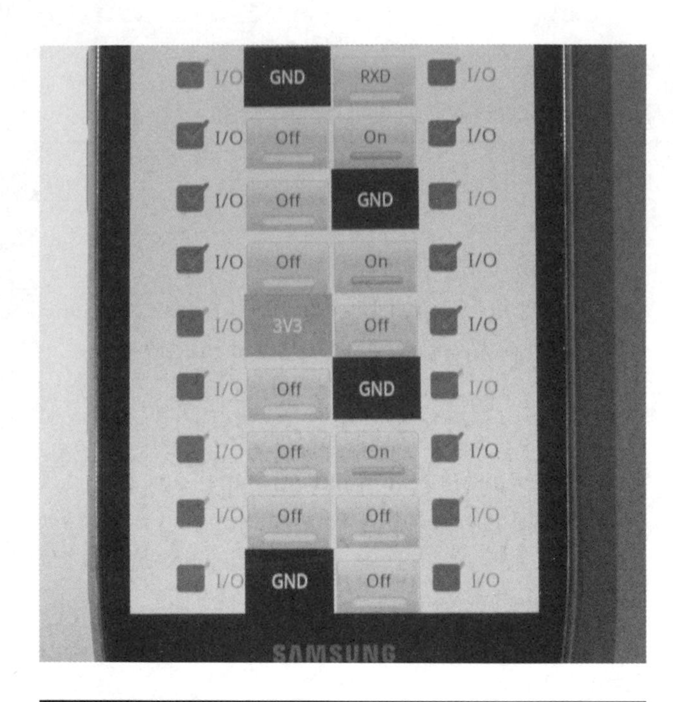

Figure 11-14 The DrGPIO control screen.

the pin box will fill in, and the label will change from "off" to "on". The pin will now be outputting a 1, or high level (3.3 V DC). Figure 11-15 is a picture of the test setup with the three LEDs turned on.

Testing the inputs is very easy. Just select the pins that you want to be used for inputs by ensuring that the check marks are highlighted and then connect those pins to the 3.3 V DC power supply. Figure 11-16 shows the phone screen with pin 18 set as an input and also connected to the 3.3 V DC supply. Pins 23 and 24 were left as outputs.

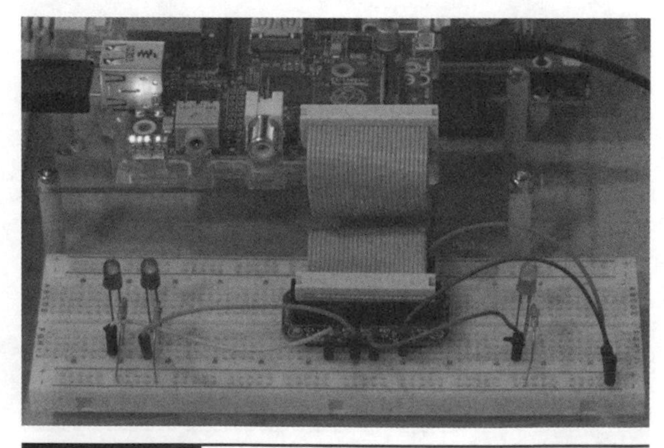

Figure 11-15 Functioning output test setup.

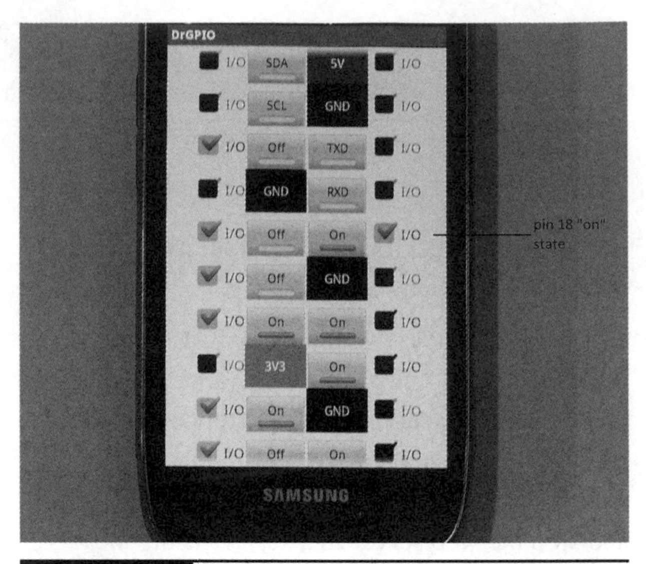

Figure 11-16 A phone input test screen.

Sharp-eyed readers may have spotted two additional pins that are set up as inputs and appear as if they are "on," yet are not connected to the 3.3 V DC. These two pins are called *floating inputs*, meaning they are unconnected and can receive stray voltage that can trigger a false reading of a high input. I deliberately left this condition in the figure to show you that you can unwittingly create a short circuit to ground by resetting a floating input GPIO pin to output without removing the ground wire.

Testing the Web Server with a Browser

The HTTP web server may also be run from any local computer's browser. Simply type in the RasPi's local IP address with port 8000 as a suffix. In my case this was:

```
http://192.168.1.21:8000
```

Figure 11-17 is a display of the resulting web page. All the configurable GPIO pins were set as outputs for this screenshot. All the functionality discussed in the Android section is available using a browser. Using a browser in this manner makes configuring GPIO pins very easy without the need to create any Python code to test basic I/O for a

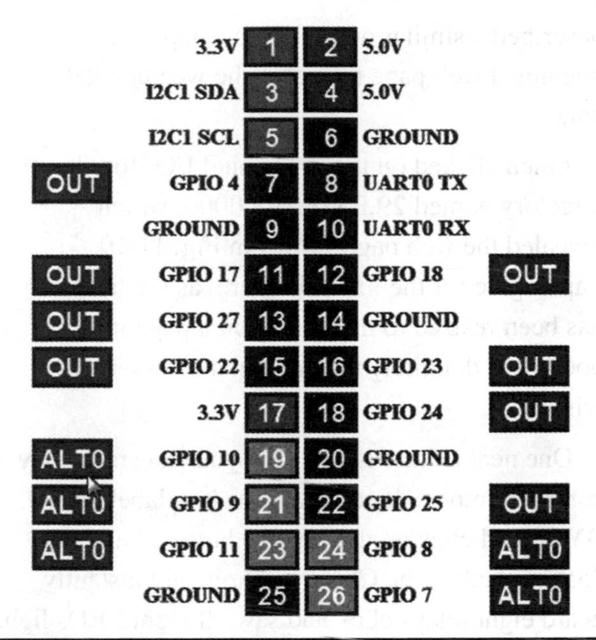

Figure 11-17 WebIOPi web page using a browser.

hardware project. I highly recommend that you consider using this approach to speed up your project development.

Internet Access

You will need your router's public IP address to be able to remotely control the RasPi's GPIO pins. Determining this address has already been discussed several times in previous chapters. You will also need to enable port forwarding on the router so that any browser or smartphone using port 8000 will automatically be forwarded to the RasPi's local IP address. I had no problem in setting this up on the Android smartphone and on a remote computer's browser.

The only issue is to remember which GPIO pins are connected to which devices. It would definitely be a problem if you wanted to turn on your home's front outdoor light but accidently started the lawn irrigation system. The WebIOPi GUI may be customized to use different widgets and labels to suit your particular application. It does require a knowledge of HTML and Javascript to accomplish this, which is well beyond this book's scope. I refer you to the WebIOPi documentation located at https://code.google.com/p/webiopi/. There is a lot of information at this site, including a discussion of *Representational State Transfer* (REST), which is the underlying technology that powers the web server. It is well worth your time to look at the content and gain an appreciation of what is happening "behind the curtains".

Remote Access to the 8 Channel I/O Board

The first step is to ensure that the owfs HTTP server is running. Type this at the command-line prompt:

```
sudo /etc/init.d/owhttpd start↵
```

Next, open a browser on a computer that is part of the local network and go to the RasPi's local IP address with port 2121 appended. In my case, this was:

```
http://192.168.1.21:2121
```

Figure 11-18 is a screenshot of the opening web page. This should look somewhat familiar, as I

OWFS on localhost:4304

directory

top	highest level	directory
81.BA0731000000	81.BA0731000000	directory
29.B82E08000000	29.B82E08000000	directory
bus.0	bus.0	directory
uncached	uncached	directory
settings	settings	directory
system	system	directory
statistics	statistics	directory
structure	structure	directory
alarm	alarm	directory

Figure 11-18 Initial web page for 8 Channel I/O board.

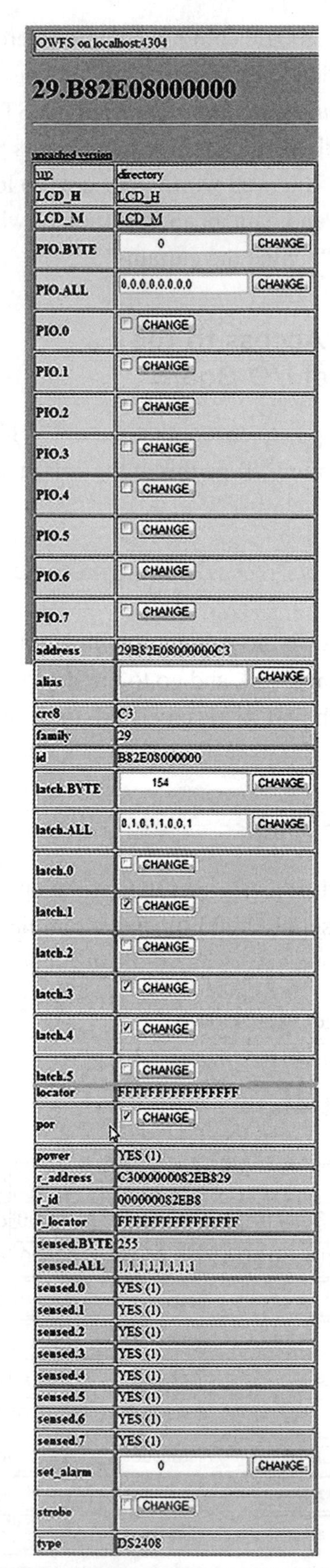

Figure 11-19 Directory 29.B82E08000000 web page.

described a similar operation in Chap. 10, when opening a web page to access the weather station data.

I then clicked on the 8-Channel I/O Board directory named 29.B82E08000000, which revealed the web page shown in Fig. 11-19. I apologize for the fine print here, as the figure has been resized to make it fit on a page in this book. It is definitely readable in a browser window.

One neat feature is that now you can enter relay control numbers directly into the box labeled PIO. BYTE and click on the Change button. I entered 255, clicked on the Change button, and instantly heard eight relays click and saw all eight LEDs light up. You should also review Fig. 11-5 to understand why this web page is so big. All the files listed in the device directory are also displayed on this web page. All the data operations discussed in the section entitled "Testing the Inputs" are available through this web page. Figure 11-20 shows the result of entering 255 in the PIO.BYTE box.

I did notice that starting the owfs web server stopped the owfs file system application and unmounted the files in the 1wire directory. I do not

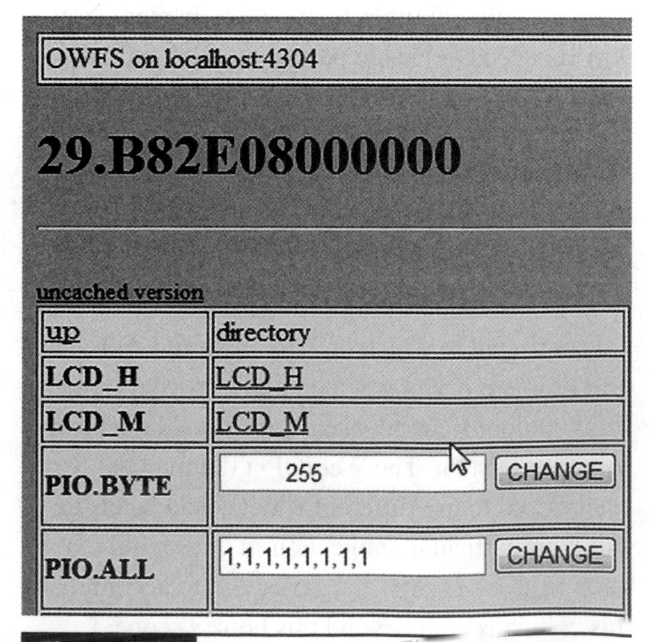

Figure 11-20 Testing the PIO.BYTE data entry.

believe this should be a real problem because you will likely be using either one or the other but not both simultaneously.

Internet access will be done in exactly the same way as discussed above except the forwarding port is 2121.

Summary

The 1-Wire 8 Channel I/O board was first examined with a detailed review of the board's key specifications. This board provides an expanded number of GPIO pins beyond the RasPi's seven.

Next, a 12-step build and configuration process that created an owfs development environment supporting the I/O board was carefully explained.

I then took a brief digression to explore Linux, FUSE and owfs. This explanation should help you understand how owfs functions.

Testing the I/O board inputs and outputs came next. I used only commands entered at the terminal prompt to exercise the board functions.

A short Python program, ow_eight.py, that controlled the I/O board through a series of output commands was demonstrated. A brief view of the Sniffer Module output was also shown. The ow_eight.py program should serve as an example of how to create more complex programs suited to unique applications.

I next showed you how to control the RasPi's GPIO pins by using an Android smartphone application named DrGPIO along with the WebIOPi web server installed on the RasPi. I also demonstrated how you could operate the same functions with the same WebIOPi web server by using a regular browser.

The chapter concluded with a demonstration on how to control the I/O board using the HTTP owfs web server from a browser.

CHAPTER 12

Robotic Car: Part 1

Introduction

This chapter is one of two that will show you how to build a RasPi controlled robotic car. I've spread the project over two chapters because of the build complexity and the software involved. The elements covered here in Part 1 include the mechanical construction and the hardware and software that control the drive system.

Developing the robotic car has been a very enjoyable experience, and the project has attracted a lot of interest among folks who are curious about what the Raspberry Pi is and how it can be used. There is nothing like a robotic car to stir up interest in this great little board. Now on to building the car!

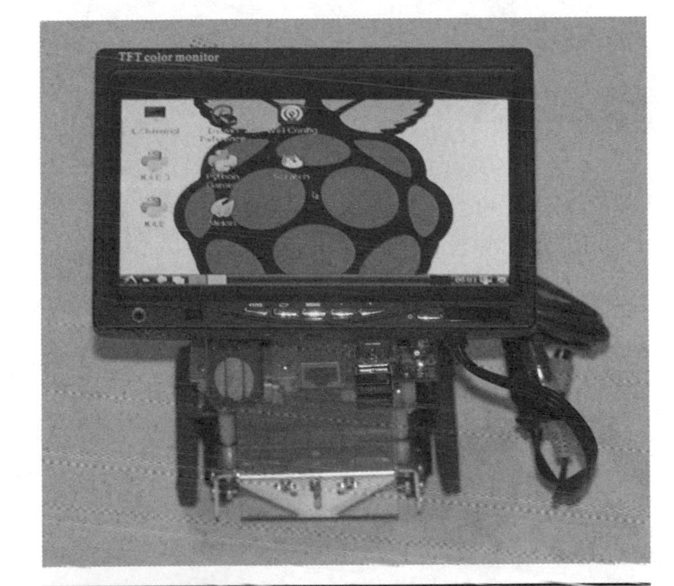

Figure 12-1 Raspberry Pi robotic car (front view).

Overview

I thought it would be a good idea to show you some pictures of the car all built and ready to run. Figures 12-1 to 12-3 show the final project with the optional *thin-film transistor* (TFT) display mounted.

And yes, that is the Wheezy Linux desktop being displayed in Fig. 12-1. I use a wireless keyboard and mouse to interact with the RasPi just as I would normally when sitting at my desk. This is truly a portable computer on wheels.

The car, however, will not be controlled by keyboard or mouse commands but will instead be handled by a Wii remote control device (Wiimote), as shown in Fig. 12-4.

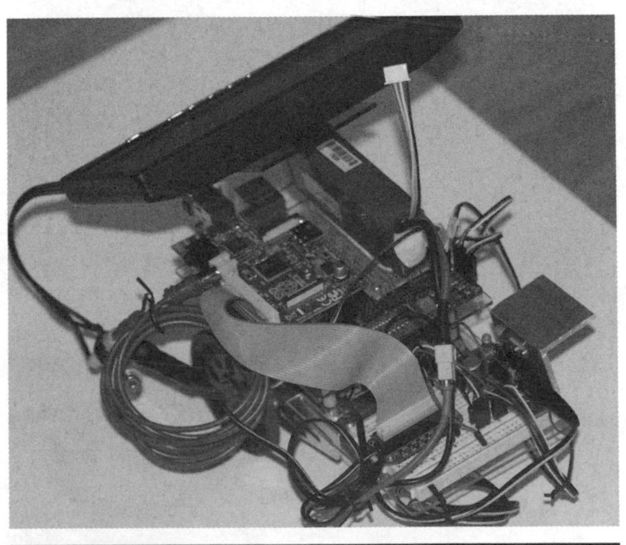

Figure 12-2 Raspberry Pi robotic car (left side view).

Figure 12-3 Raspberry Pi robotic car (right side view).

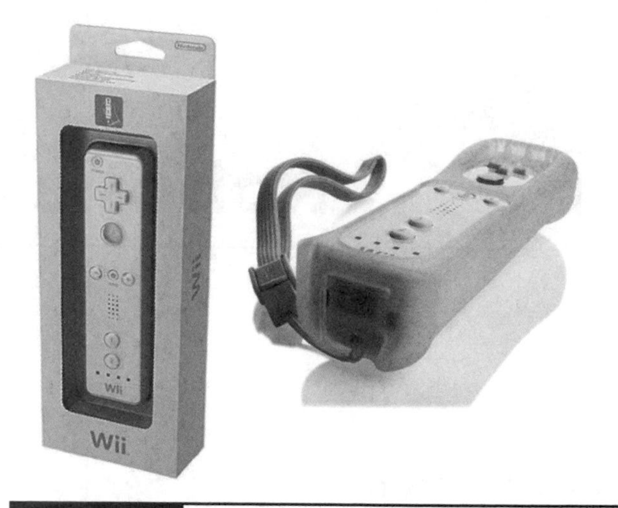

Figure 12-4 Wiimote.

The Wiimote uses *Bluetooth*, a wireless technology for data transfer over short distances, to communicate with the RasPi, which will be discussed in Chap. 13.

Chassis and Drive Motors

The car uses a simple tripod configuration of two powered wheels and a third one as a free rolling ball to provide stability and a very tight turning radius. All the chassis, drive, and power supply components are listed in Table 12-1 and are available from www.parallax.com

A servo control board is also required for this project. It is the 16-Channel 12-bit *Pulse Width Modulation* (PWM) /Servo Driver I²C Interface board available from Adafruit Industries, part number 815.

The Boe Bot chassis is made of 1.5 mm aluminum and is shown in Fig. 12-5.

A complete drawing showing all the dimensions is available on the Parallax website for readers who are able and willing to make their own chassis. Just use the part number to look it up.

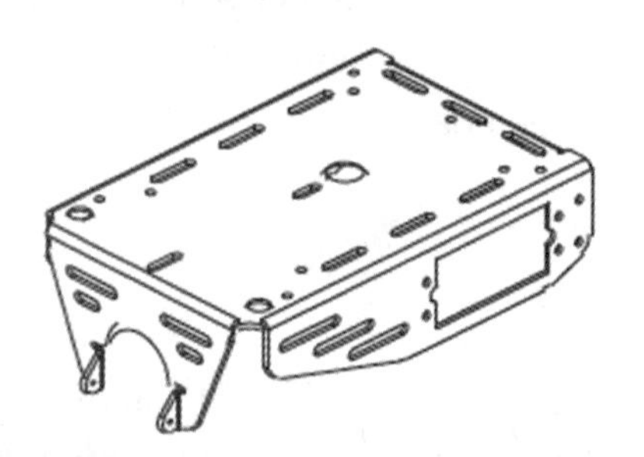

Figure 12-5 Car chassis.

Table 12-1 Chassis, Drive, and Power Supply Components

Quantity	Description	Parallax Part Number	Price (USD) (each)
1	Boe Bot chassis	700-00022	24.99
2	Continuous rotation servo (Futaba)	900-00008	12.99
2	Boe Bot wheels and tires	28109	3.50
1	Polyethylene ball	700-00009	3.95
1	Cotter pin	700-00023	0.30
1	Li-ion Power Pack Charger	28986	52.96
2	Li-ion high capacity cell	28987	7.09

Figure 12-6 Continuous rotation servo.

The continuous rotation servos provide the driving force for the wheels. Figure 12-6 shows a servo, and Fig. 12-7 shows a wheel kit. Each wheel is attached to the drive servo with a 0.25-inch screw that is also provided with each servo.

The polyethylene ball is shown in Fig. 12-8, and it is held in place with the cotter pin specified in the parts list.

Figure 12-9 illustrates the bottom side of all the assembled chassis parts. The servos are mounted through the cutouts in the aluminum chassis with 0.5-inch 4-40 machine screws and nuts. This

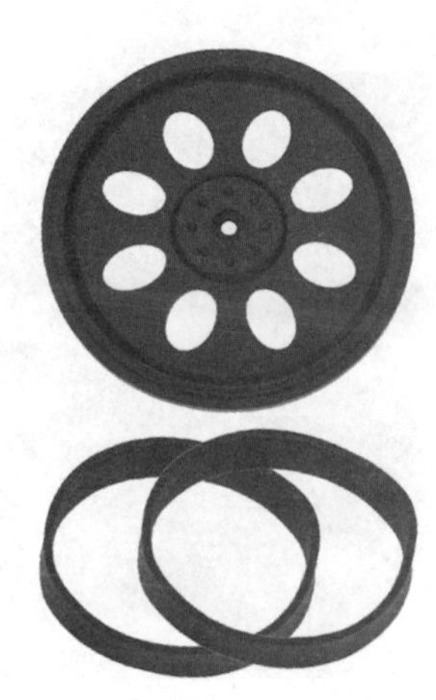

Figure 12-7 Wheel kit.

Figure 12-8 Polyethylene ball.

generic hardware should be readily available at local hardware or home supply stores.

You can also see a hole conveniently placed in the middle of the chassis top that is used to thread the servo control wires to the servo control board. The chassis assembly should be a very quick and easy procedure if you have all the kit parts available. Parallax has Boe Bot assembly instructions available on its website if you find that you need additional guidance.

Mounting the power source is the next step in the assembly process. I chose to use a Parallax Li-ion power charger board because it has ample capability to power a robot car and it just happens that its form factor precisely matches the Boe Bot

Figure 12-9 Bottom view of the assembled car chassis.

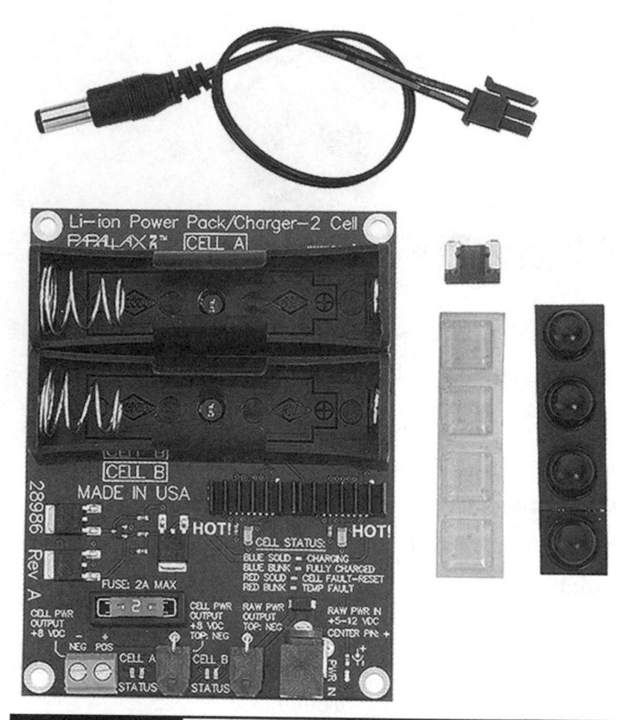

Figure 12-10 Li-ion charger board.

chassis including the mounting holes. Figure 12-10 shows the Li-ion charger board.

This board uses two high capacity Li-ion cells rated at 3.7 V DC and 2600 mAh. One cell is shown in Fig. 12-11. These cells conform to the 18650 form factor, which is a bit like an AA battery on steroids. Each cell also contains a protection circuit to help prevent overcharging, which could lead to a serious situation if the cell was left unprotected. Nonetheless, these cells should be charged *only* with an appropriate charger specifically designed for them, which is the main reason why I incorporated the complete charger into the car.

NOTE **The U.S. Postal Service will no longer ship items that contain lithium batteries to international destinations, including APO and FPO addresses.**

Figure 12-11 Li-ion cell.

Figure 12-12 Charger board Sketchup mounting diagram.

The charger board and all the other boards, including the RasPi, are mounted in a stacked "sandwich" arrangement. Please note that you will need to assemble and disassemble the stack several times as you build the car. This is unavoidable because you will have to custom fit the parts to the various mounting boards. Figure 12-12 is a *Sketchup* (3D modeling software once owned by Google but now owned by Trimble) representation of the beginning of the stack with the charger board mounted on four 0.25-inch OD nylon spacers.

Four 1-inch nylon spacers are mounted on top of the board to provide sufficient room between the top of the charger board and the bottom of the Lexan mounting board. Figure 12-13 shows a portion of the actual mounted charger board so that you will gain an understanding of this arrangement.

Figure 12-13 The charger board mounting

I used a 2-mm 4- × 4.5-inch Lexan board as a mounting platform to support the RasPi and the servo control board. Figure 12-14 shows this board along with the charger board and the nylon spacers. The whole stack is held together by four 2.25-inch-long 6-32 machine screws that go through the Lexan board, hollow spacers, and the Li-ion charger board. I used a 6-32 tap to prethread the four matching holes on the Boe Bot chassis to make the installation a bit easier. You can choose to do this or to simply use a 6-32 nut and washer to lock down the screw. I believe you will need to slightly enlarge the chassis holes for clearance if you use screws and nuts.

I would also recommend using the charger board as a drill template for the Lexan mounting board. Just ensure that the Lexan board is mounted to have the 4-inch dimension between the wheels and the one edge flush with the charger board, as shown in Fig. 12-14. It is important to ensure that the optional monitor support can be installed without interference.

Next, disassemble the Lexan board to mount the RasPi and servo control board. I used two 0.75-inch 4-40 screws and nuts along with two 0.5-inch nylon spacers to mount the RasPi to the Lexan board. Use the two mounting holes on the RasPi as drill guides for locating the holes in the Lexan board. Ensure that the RasPi board edge lines up

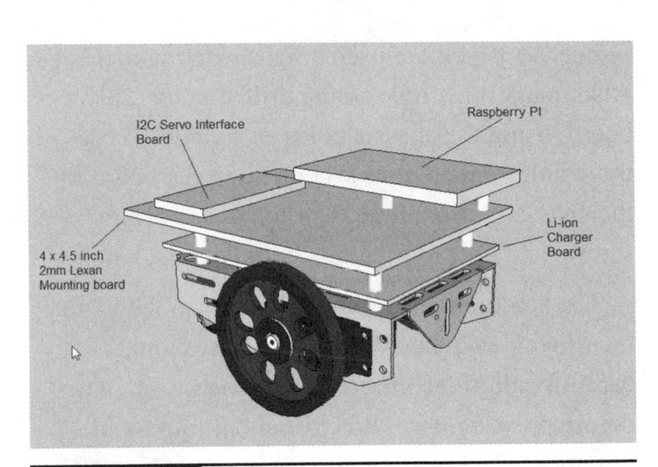

Figure 12-15 RasPi and servo boards mounted.

with the Lexan board edge, as shown in Fig. 12-15. The servo control board is mounted as shown in the diagram except that there are four mounting holes for that board. In addition, I used nylon washers to space the servo board from the Lexan board. The mounting holes are very tight, so you will need four very tiny machine screws and nuts.

Another Lexan mounting board is needed to support the solderless breadboard used for the Pi Cobbler and other items that can be seen in Figs. 12-2 and 12-3. This additional Lexan board is 3.5 × 4 inches and is mounted at the back of the car directly to the chassis, as shown in Fig. 12-16. You will need to notch out the board to provide clearance for two spacers supporting the charger board. I drew the cutouts on the board's Lexan

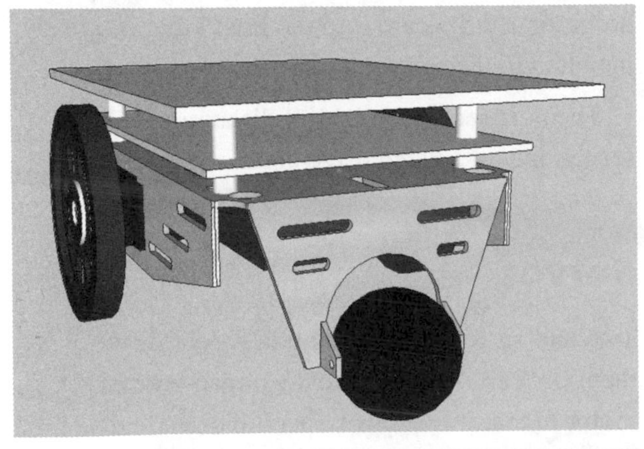

Figure 12-14 Lexan and charger boards mounting diagram.

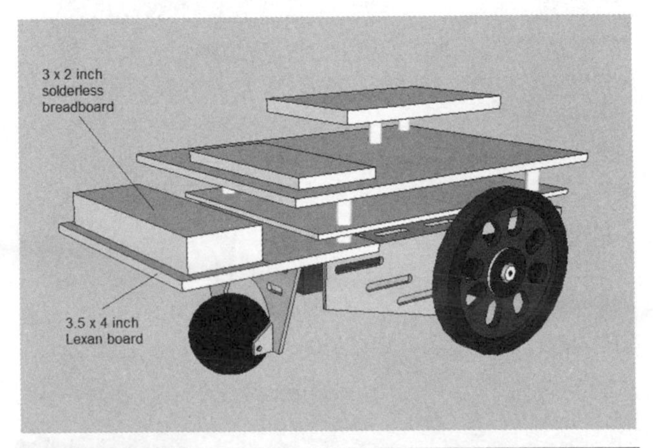

Figure 12-16 Lexan board supporting the solderless breadboard.

protective paper and used a jig saw to make them. Also, mark three holes to be drilled in the Lexan board, using the chassis holes as a guide. These three holes are obvious when you fit the board to the chassis. I used three 0.5-inch 4-40 machine screws and nuts to hold this board to the chassis.

The solderless breadboard mounts to the Lexan board with an adhesive backing. You should carefully align the breadboard to the Lexan plate before pressing it on. It is just about impossible to remove the breadboard once it is attached.

The remaining mechanical piece is the Lexan support plate for the portable monitor, if you should so choose to install it. It is a fairly simple piece to install with the only tricky part being cutting clearance holes for the RasPi USB ports and the additional LiPo battery that is used to power the monitor. Figure 12-17 is a dimensioned sketch of the Lexan support that I used. Your support piece will likely vary a little from mine, since your RasPi placement may also vary from my installation.

The support is made out of one 2.5- × 8-inch Lexan sheet bent at a 45° angle in the middle of the long dimension. I used a hot-air gun to soften the Lexan so that it could be easily shaped. The support is attached to the chassis front with three 0.5-inch 4-40 screws and nuts, as shown in Fig. 12-18.

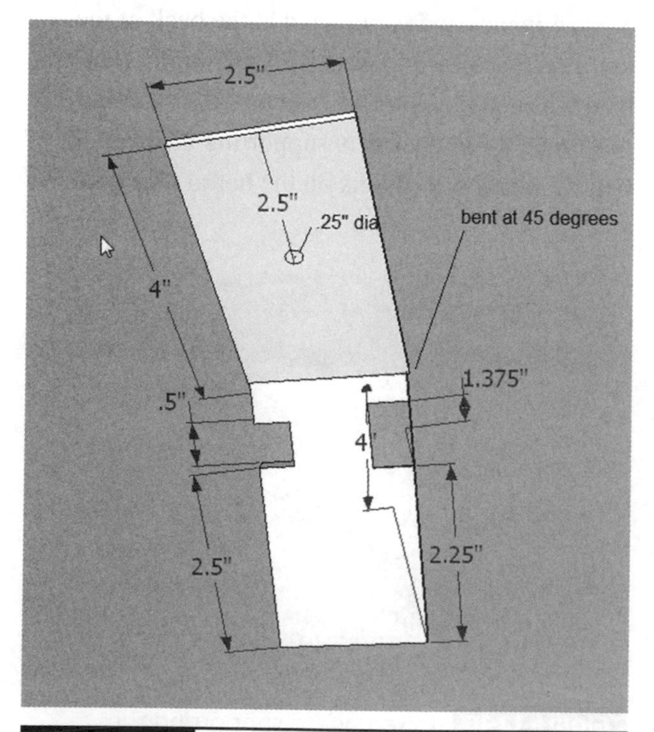

Figure 12-17 Lexan monitor support piece.

Figure 12-18 Monitor support plate attachment.

This last step completes the robot car mechanical construction phase. The electrical connections will be shown in Chap. 13.

You may, however, have an immediate issue if you have only one RasPi. You need to program the RasPi to run the car, which is very hard to do when it is installed in the car. I would remove it from the car and program it as you would normally and then reinstall it when everything is all set up. Obviously, if you have a second RasPi, you can use that one to do all the programming and then put the programmed SD card into the RasPi that is already mounted in the car.

The next sections are brief discussions on the servos and associated drive components.

Servos

Two analog continuous rotation servos drive the robot car. A servo differs from an ordinary motor in that it requires a specific digital pulse train to operate. Regular or non-continuous servos are designed to rotate a shaft a certain

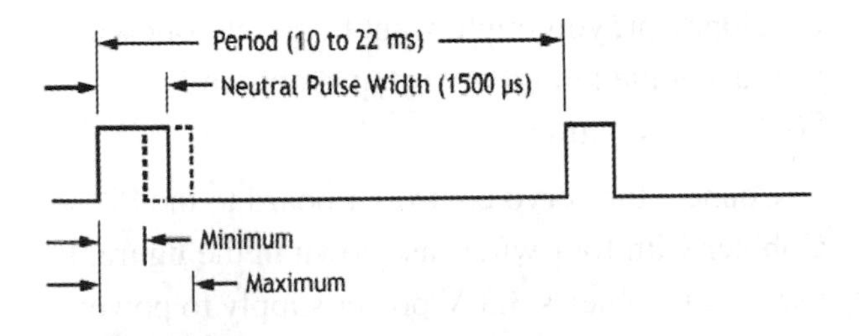

Figure 12-19 Typical PWM signal.

number of degrees based upon the *pulse-width modulation* (PWM) of the control signal. Continuous rotation servos are designed to maintain a certain angular velocity based upon the PWM control signal. In either case, the standard PWM signal frequency is set at a nominal 50Hz and operates at a nominal 5 V DC for a high level and 0 V DC for a low level. Figure 12-19 shows a typical PWM signal.

The neutral position, where neither a position nor a velocity is being commanded, is set to be a 1500 μs pulse width. The servo controller will increase or decrease the pulse width up to 500 μs depending upon the new desired position or velocity. The servo motor contains an electronic circuit that converts the PWM pulse train to the appropriate internal motor control signals that move the servo to the commanded position or angular velocity.

Table 12-2 shows some of the specifications for the Parallax servos used in this robot car.

Table 12-2 Servo Specifications

Specification	Value
Servo manufacturer	Futaba
RPM range	0 to 50
Operating voltage range	4 to 6 V DC
Max torque	38 oz-in @ 6 V DC
Weight	1.5 oz (42.5 g)
Mounting	Four mounting screws

I²C and the Servo Controller

A servo controller board is required for this project because the RasPi only has one channel that can output PWM signals and the car needs two, one for each motor. I used the 16-channel, 12-bit PWM/Servo Driver I²C Interface Board available from Adafruit Industries. It is a bit of overkill for this application, but it is flawless in operation and very easy to set up.

I²C is another one of the serial protocols that was first introduced in Chap. 1. The servo control board uses I²C to communicate with the RasPi. However, the RasPi must first be configured to operate with the I²C protocol, which is not built into the Wheezy distribution (as of the date of this writing). It is part of the Occidentalis v.2, if you choose to use that version, and is not very hard to setup Wheezy with the I²C protocol.

At the command line prompt, enter the following:

```
sudo nano /etc/modules ↵
```

Add the next two lines to the end of the file.

```
i2c-bcm2708
i2c-dev
```

Reboot the RasPi for the changes to take effect.

```
sudo reboot ↵
```

Enter the following commands after booting the RasPi:

```
sudo apt-get update ↵
sudo apt-get upgrade ↵
```

(Be patient, this could take a while.)

These commands ensure that the Wheezy distribution is current with all the latest changes. Next, enter:

```
sudo apt-get install python-smbus ↵
sudo apt-get install i2c-tools ↵
```

These commands load some useful applications that, among other things, allow you to check the address of attached I²C modules. The i2c-tools package contains the i2cdetect application that will display all the I²C devices connected on the bus. Using it will confirm that the servo controller is being recognized by the RasPi.

The python-smbus package contains all the software needed to access I²C devices from Python. This library installation is critical to ensure that the robot control program works.

The next step may not be required. It depends upon the Wheezy distribution that you are running. Check if there is a file name /etc/modprobe.d/raspi_blacklist.conf installed. You do nothing if it doesn't exist; otherwise comment out these lines:

```
blacklist spi-bcm2708
blacklist i2c-bcm2708
```

Enter the command below to edit the file, and put a "#" symbol in front of the lines shown above:

```
sudo nano /etc/modprobe.d/raspi_
blacklist.cong ↵
```

I²C Initial Test

I would defer testing the I²C bus if you have only one RasPi and it is already installed in the car. However, if you are using a second RasPi for development, you might want to do the following test, using the temporary setup shown in Fig. 12-20, I²C test diagram.

Connect the servo controller board to the Pi Cobbler with four wires, as shown in the figure. I used the Cobbler's 3.3-V power supply to power the servo controller board to ensure that 3.3 V was the maximum voltage that could appear on the I²C bus lines. However, I do believe that the controller itself limits the voltage to 3.3 V even if it is powered by a 5-V source, but I could not confirm it from any data sheet. Better safe than sorry, as the old adage goes.

CAUTION **The servo control board has a two-screw terminal strip that routes power directly to the servos. This power source takes up to 6 V DC and is separate and distinct from the V_{CC} supply that powers the board logic. Do not connect the servo supply to the V_{CC} supply, or you will destroy the servo control board.**

Enter the following at a terminal control prompt:

```
sudo i2c-detect –y 1 ↵
```

NOTE **If you are using an early 256 MB RasPi model B, change the 1 to a 0 in the above command.**

Figure 12-21 shows the result of this command. In the figure you will see that two I²C addresses

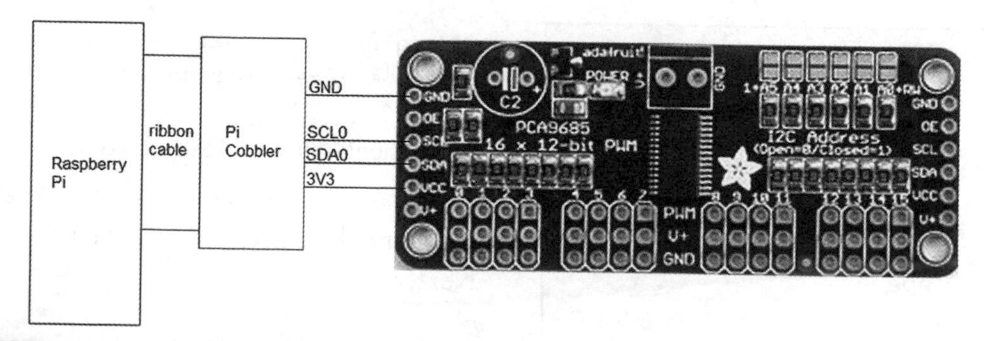

Figure 12-20 Temporary I²C test diagram.

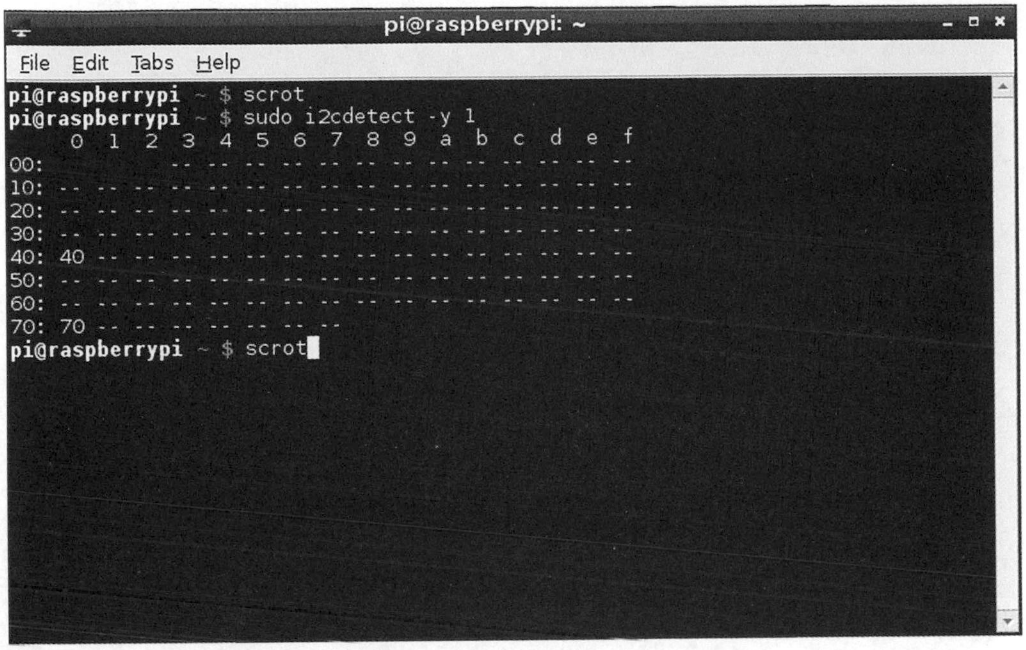

```
pi@raspberrypi: ~
File Edit Tabs Help
pi@raspberrypi ~ $ scrot
pi@raspberrypi ~ $ sudo i2cdetect -y 1
     0  1  2  3  4  5  6  7  8  9  a  b  c  d  e  f
00:          -- -- -- -- -- -- -- -- -- -- -- -- --
10: -- -- -- -- -- -- -- -- -- -- -- -- -- -- -- --
20: -- -- -- -- -- -- -- -- -- -- -- -- -- -- -- --
30: -- -- -- -- -- -- -- -- -- -- -- -- -- -- -- --
40: 40 -- -- -- -- -- -- -- -- -- -- -- -- -- -- --
50: -- -- -- -- -- -- -- -- -- -- -- -- -- -- -- --
60: -- -- -- -- -- -- -- -- -- -- -- -- -- -- -- --
70: 70 -- -- -- -- -- -- --
pi@raspberrypi ~ $ scrot
```

Figure 12-21 Detecting the I²C servo controller.

were detected, one at 0 ×40 and the other at 0 ×70. The 0 ×40 is the default slave address preset into the PCA9685 chip that is the logic controller used in the servo board. I believe that 0 ×70 is another controller address that is generated during a power-on reset. In any case, 0 ×40 is the only address that will be used in the Python program. This address may be changed in hardware by connecting solder bridges located on the board. These bridges are easily seen on the upper right-hand side of the board, as shown in Fig. 12-20. Board addresses would have to be changed, if multiple servo controllers were used on the I²C bus, to prevent addressing conflicts.

This chapter does not have a summary section, as the project is not yet complete. The next chapter shows how to wire the car and program it so that it can be controlled by a Wiimote.

CHAPTER 13

Robotic Car: Part 2

Introduction

This is the second of two chapters that will show you how to complete the RasPi-controlled robotic car, a project that was started in Chap. 12. This chapter deals with the electrical wiring and most of the programming required to control the car.

Robotic Car Block Diagram

Figure 13-1 is a block diagram showing all of the robotic car electrical components.

The RasPi is powered from the Li-ion charger board; however, that board provides only an 8-V DC output. The voltage is converted to 5 V DC by a small regulator board that uses an LM7805 three terminal chip with some auxiliary capacitors. Figure 13-2 shows the regulator schematic.

The 5-V output is connected to the Pi Cobbler's 5-V pin 0. Technically, this arrangement is known as *back feeding* because the 5-V 0 pin is normally an output, not an input. However, it makes no difference electrically if you supply or take off power from a parallel-connected power bus. By following this approach, you avoid having to hack up a micro USB connector and cable through which the RasPi is normally powered.

Shown on the block diagram is another 5-V DC regulator that provides power to the servo motor input on the 16-channel servo control board (now simply referred to as the servo board). You might wonder why I did this instead of paralleling the output already connected to the RasPi. The answer is that I wanted to avoid the situation in which a brief high-current demand from the servo motors would cause the regulator voltage output to drop below the minimally acceptable voltage for the RasPi, thus causing it to stop running.

The logic on the servo board is fed from the Pi Cobbler's 3-V pin 3 for reasons already stated in Chap. 12.

The TFT monitor requires 270 mA at 12 V DC to operate. This requirement necessitated that an additional *lithium-ion polymer* (Li-Po) battery be added for the sole purpose of powering the monitor. The Li-Po battery is normally used in a radio-controlled aircraft application where it can supply fairly high current for short time bursts. However, it fits this situation quite well, being able to power the monitor for about eight hours, given its 2200-mA/h rating. Figure 13-3 shows this battery.

CAUTION **The Li-Po battery must be recharged only with an approved charger designed for it. Under *no* circumstances should you use an automotive 12-volt charger with this battery. Bad things, such as a fire, will happen!**

You may also have noticed a Velcro™ strip attached to the underside of the battery. I used this strip as a convenient way to mount the battery and to allow it to be quickly removed for a recharge.

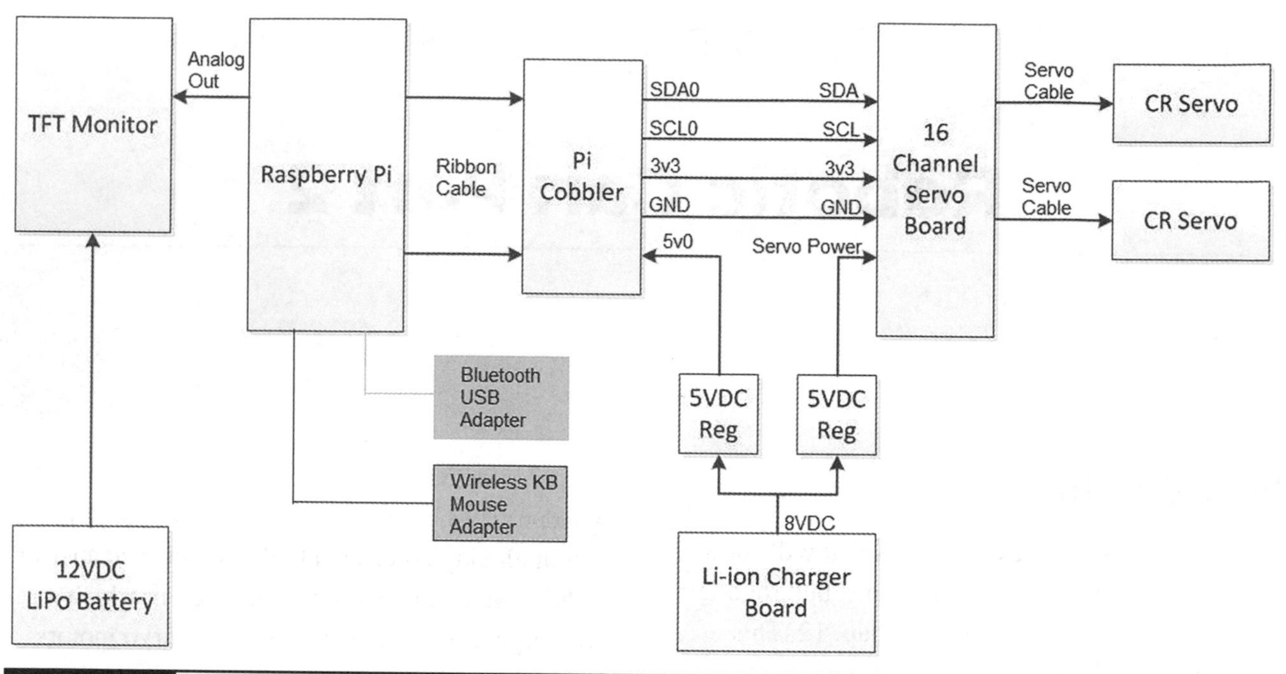

Figure 13-1 Block diagram of the robotic car electrical components.

There are two Bluetooth micro USB adapters plugged into the RasPi's USB slots. One is for the Wiimote, and the other is for the wireless keyboard and mouse unit that I use with the car. Figure 13-4 shows this Logitech model K400r keyboard and mouse combination. I have found it to be extremely reliable and very moderately priced.

The servo motors connect to the servo board by means of the three lead cables already attached to the motors. Each cable plugs directly into one of the 16 three pin connectors. I used connector numbers 0 and 1 for the right- and left-hand motors, respectively. It doesn't matter which servo connectors are used as long as the software is appropriately modified to reflect the actual connectors. There is a more or less standard color code used for servo cables as follows:

red = power

white = signal

black = ground

I have one last item to mention regarding the wiring: the Li-ion charger board's power socket

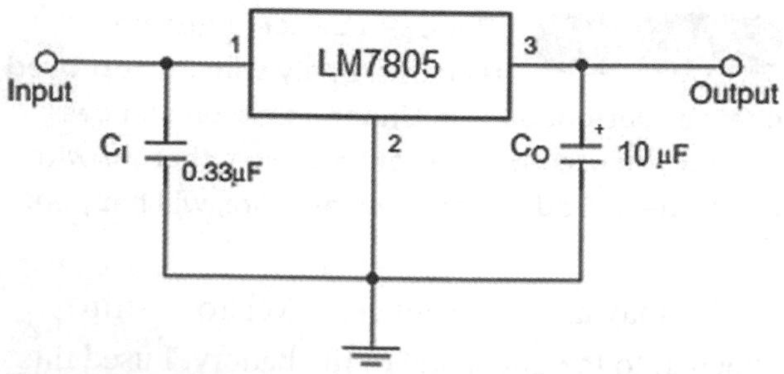

Figure 13-2 5-V DC regulator schematic.

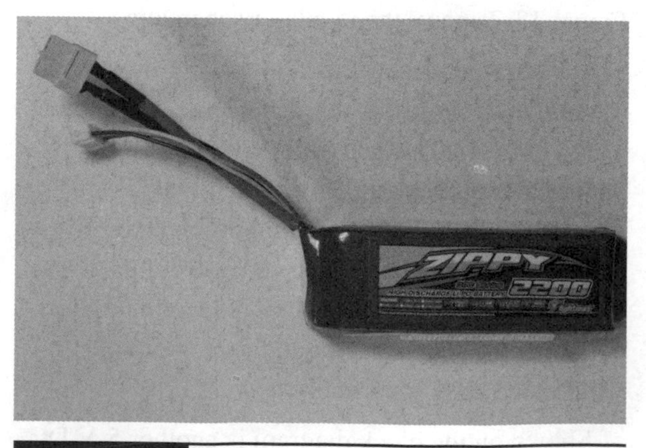

Figure 13-3 The Li-Po battery for the TFT monitor.

Figure 13-4 Logitech K400r wireless keyboard/mouse.

should be accessible so that it can be recharged. Figure 13-5 shows such an arrangement on the prototype robot car.

Notice also that I had to remove the RasPi 5-V regulator board from the breadboard in order to provide clearance for the power cord. I soldered two pins to the regulator so that the board plugs directly into the 5-V and GND strips that are the outermost strips on the breadboard. It makes it very convenient to have a pluggable regulator for use with a breadboard.

I²C Software

The servo board communicates with the RasPi using the I²C protocol. The I²C RasPi setup, which has already been discussed in Chap. 12, must be completed before proceeding with the remaining software installations.

Figure 13-5 The wall wart power cord plugged into the Li-ion charger board.

Bluetooth Software

The first step in setting up the car to use Bluetooth is to plug in the Bluetooth adapter. Just make sure the RasPi is not running. Figure 13-6 shows the V4 Targus Bluetooth adapter that I used in this project. I am sure that similar type adapters will function equally as well, but just ensure that you choose a high-speed class-4 device.

Now, start the RasPi and check that the Wheezy distribution has recognized the Bluetooth adapter. Enter the following at a command line prompt:

```
lsusb ↵
```

The listing in Fig. 13-7 shows that the adapter has been found and an appropriate driver loaded for it. The Targus adapter apparently uses a Broadcom

Figure 13-6 Targus Bluetooth adapter.

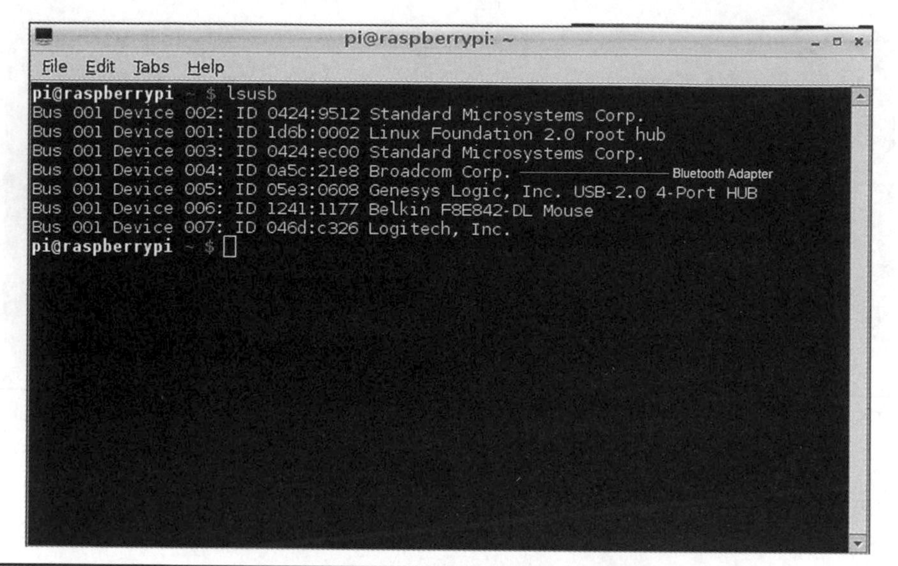

Figure 13-7 lsusb listing.

chip, from the same fine company that makes the RasPi core processor.

You next need to download and install the bluetooth software package. Enter the following at a command line prompt:

```
sudo apt-get update ↵
sudo apt-get install bluetooth ↵
```

(Be patient, this takes a while.)

Enter the following to check if the Bluetooth service is running:

```
sudo service bluetooth status ↵
```

Figure 13-8 shows the status for an up-and-running service.

If the status shows that it is not running, enter this command:

```
sudo /etc/init.d/bluetooth start ↵
```

Reenter the above status command to confirm that the service is now running.

Enter the following command to check on the adapter's address:

```
sudo hcitool dev ↵
```

Figure 13-9 shows the resulting screen display of the address of the Bluetooth adapter. The address is not really needed, but it does check that the hcitool application was loaded with the Bluetooth package.

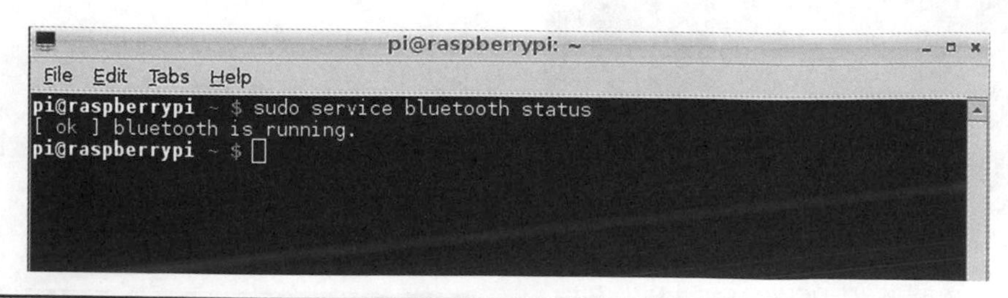

Figure 13-8 Bluetooth status.

```
pi@raspberrypi ~ $ hcitool dev
Devices:
        hci0    00:19:0E:11:16:DA
```

Figure 13-9 Bluetooth adapter address.

The stage is now set to start scanning for Bluetooth-enabled devices. Gather up your Wiimote and ensure that it has fresh batteries installed. Enter the following command to start the scan:

```
sudo hcitool scan ↵
```

You will see "`Scanning ...`" displayed on the console screen. Simultaneously press the 1 and 2 buttons on the Wiimote to put it into the discovery mode. Don't fret if it takes several tries to discover the remote. The screen will show a time-out if the RasPi doesn't discover the remote within 5 to 10 seconds. Just try again. Eventually you will be rewarded with a display, as shown in Fig. 13-10.

Notice that all Bluetooth-enabled devices in the immediate vicinity are discovered. In Fig. 13-10, both the Wiimote (Nintendo RVL-CNT-01) and an HP wireless printer were discovered. Don't be concerned with conflicting Bluetooth devices because the protocol is quite capable of keeping track of which device needs to be addressed by an application.

Another software package is needed to enable a Python program to interface to a Bluetooth device. This package is named *cwiid* and is downloaded and installed by entering:

```
sudo apt-get install python-cwiid ↵
```

This package contains all the necessary applications that allow the Wiimote to interface

```
pi@raspberrypi ~ $ hcitool scan
Scanning ...
        00:21:BD:9A:01:34       Nintendo RVL-CNT-01
        F4:CE:46:0D:A9:4B       Photosmart Prem-Web C309n-s
```

Figure 13-10 Wiimote discovered.

with a user-created Python program. I would also like to extend my thanks to Brian Hensley (www.brianhensley.net) for his fine tutorial on using the Wiimote with the RasPi. It was invaluable in helping me to create the following programs.

However, before discussing an actual program, I need to show you how the Wiimote buttons are set with regard to *state values*, which are those numbers sent to the RasPi, indicating their corresponding buttons have been pushed. Figure 13-11 is a Wiimote with all the corresponding state values superimposed on the image.

For instance, if the forward button of the four-way switch is pressed, a value of 2048 will be transmitted via the Bluetooth connection to the RasPi. Likewise, if the "–" button is pressed, a value of 16 is sent. The values are also cumulative in case of combination button presses. For example, simultaneously pressing the 1 button and the home button will cause a value of 130 to be sent: the sum of the home button value of 128 and

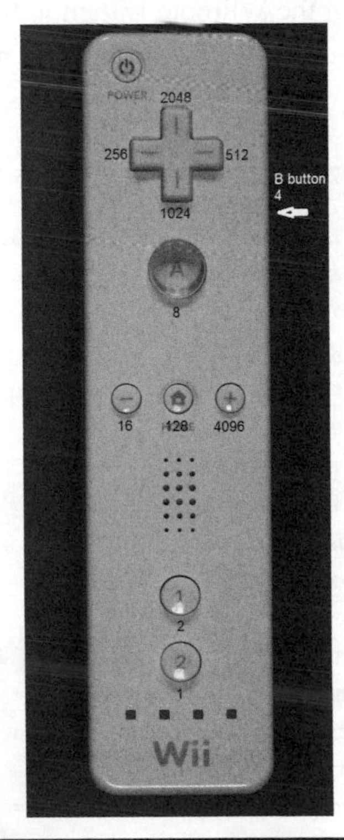

Figure 13-11 Wiimote with state values.

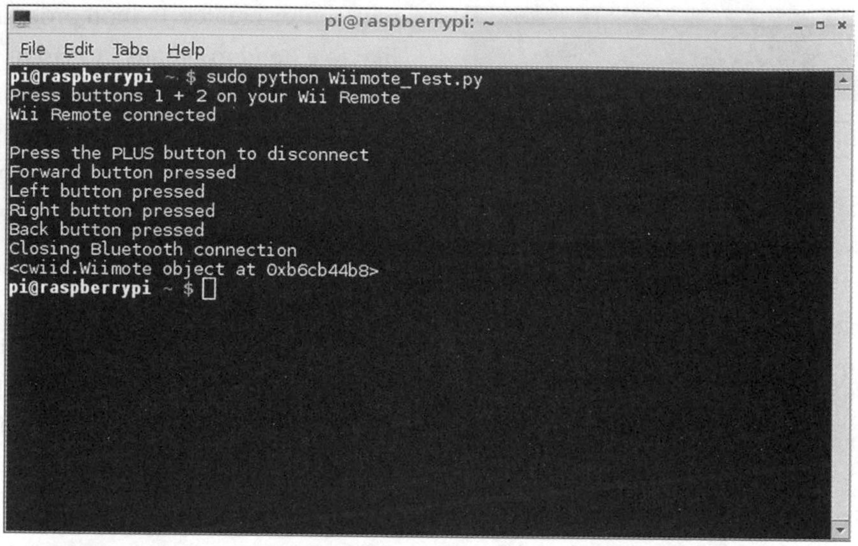

```
pi@raspberrypi: ~
File Edit Tabs Help
pi@raspberrypi ~ $ sudo python Wiimote_Test.py
Press buttons 1 + 2 on your Wii Remote
Wii Remote connected

Press the PLUS button to disconnect
Forward button pressed
Left button pressed
Right button pressed
Back button pressed
Closing Bluetooth connection
<cwiid.Wiimote object at 0xb6cb44b8>
pi@raspberrypi ~ $
```

Figure 13-12 Wiimote_Test.py program results.

the 1 button value of 2. The application becomes a matter of decoding the values received and figuring out what actions to take based upon those values.

I created a simple test program named *Wiimote_Test.py* that establishes a Bluetooth connection between the car and the Wiimote. The four-way switch on the Wiimote is then used to send different values, depending upon which switch is pressed. Enter the following to run this program:

```
sudo python Wiimote_Test.py ↵
```

Figure 13-12 shows the result obtained by first establishing a connection and then pressing all four switches.

The Python code for Wiimote_Test.py is available from the book's website, www.mhprofessional.com/raspi, and is also listed below:

Wiimote_Test.py

```
#!/usr/bin/env python
import time
import cwiid
print ‘Press buttons 1 + 2 on your Wii Remote’
time.sleep(3)
wm = cwiid.Wiimote()
print ‘Wii Remote connected’
print ‘\nPress the PLUS button to disconnect’
time.sleep(1)
wm.rpt_mode = cwiid.RPT_BTN
while (wm.state[‘buttons’] < 4096):
     if wm.state[‘buttons’] == 2048:
          print(‘Forward button pressed’)
          time.sleep(1)
     if wm.state[‘buttons’] == 1024:
          print(‘Back button pressed’)
          time.sleep(1)
```

```
    if wm.state['buttons'] == 512:
        print('Right button pressed')
        time.sleep(1)
    if wm.state['buttons'] == 256:
        print('Left button pressed')
        time.sleep(1)
print 'Closing Bluetooth connection'
time.sleep(1)
exit(wm)
```

The expression "`wm = cwiid.Wiimote()`" is all that it takes to instantiate an object that logically represents the Wiimote. The object is then placed in the proper state to report its state value by the program line "`wm.rpt_mode = cwiid.RPT_BTN`." From then on, all that needs to be done to retrieve the state value is to use the expression "`wm.state['buttons']`."

The Wiimote disconnects, and the program stops when the + button is pushed. The expression "`while (wm.state['buttons'] < 4096):`" establishes a loop and waits for the state value to exceed 4096 to stop the looping. Using a high value for this state change precludes any unintended program stoppage due to combination button presses, which is something to keep in mind when you start to create your own programs.

Robot Car Program

The program requirements for controlling the car were deliberately made very simple. The car was to go forward when the forward button on the four-way Wiimote switch was pressed. It was to turn right when the right button was pressed and turn left when the left button was pressed. It was to stop when the back button was pressed. Obviously, much more sophisticated control commands could be devised, but this simple set was deemed adequate to prove out the basic car functionality.

The program follows the same state variable detection and corresponding action as was demonstrated in the test program, only this time the commands are being sent to the servo control board, which in turn, controls the servo motors. This program would not function except for the brilliant PWM servo library provided by Adafruit Industries. This library enables PWM commands in Python that directly control the servo board and indirectly control the servo motors.

The only tricky part in developing this program was determining the PWM parameters that set the servo motors rotating at the correct angular velocity and direction. These are shown in the code as constants:

```
servoRightFwd = 300
servoLeftFwd = 2400
```

You will have to experiment with different values if you desire to have the robot go a bit faster. I know that these values make the robot travel in a straight line when so commanded.

The program is named *Robot_Car.py* and is available from the book's website. The program listing along with some amplifying comments can be found on page 172.

The program line "`pwm = PWM(0x40, debug = True`" sets up a PWM object that is at I^2C address 0×40. This is how multiple I^2C devices can be configured when they all are attached on the same bus lines.

The program segment:

```
if wm.state['buttons'] == 2048:
    pwm.setPWM(0, 0, servoRightFwd)
    pwm.setPWM(1, 0, servoLeftFwd)
```

Robot_Car.py

```
#!/usr/bin/env python
from Adafruit_PWM_Servo_Driver import PWM
import time
import cwiid
pwm = PWM(0x40, debug=True)
servoRightFwd = 300
servoLeftFwd = 2400
servoRightRev = 0
servoLeftRev = 0
pwm.setPWMFreq(60)
print 'Press buttons 1 + 2 on your Wii Remote'
time.sleep(1)
wm = cwiid.Wiimote()
print 'Wii Remote connected'
print '\nPress the PLUS button to disconnect'
time.sleep(1)
wm.rpt_mode = cwiid.RPT_BTN
while (wm.state['buttons'] < 4096):
    if wm.state['buttons'] == 2048:
        pwm.setPWM(0, 0, servoRightFwd)
        pwm.setPWM(1, 0, servoLeftFwd)
        time.sleep(1)
    if wm.state['buttons'] == 1024:
        pwm.setPWM(0, 0, servoRightRev)
        pwm.setPWM(1, 0, servoLeftRev)
        time.sleep(1)
    if wm.state['buttons'] == 512:
        pwm.setPWM(0, 0, servoRightFwd)
        pwm.setPWM(1, 0, 0)
        time.sleep(1)
    if wm.state['buttons'] == 256:
        pwm.setPWM(0, 0, 0)
        pwm.setPWM(1, 0, servoLeftFwd)
        time.sleep(1)
print 'Closing Bluetooth connection'
time.sleep(1)
exit(wm)
```

drives the car forward when the forward button is pressed. You just need to press the back button to stop this motion. The same holds true for turning right or left.

Operating the Robotic Car

The car is essentially a peripheral for the RasPi, which means that you must start the computer as you would normally do and simply enter the command to run the Robot_Car.py program as shown below:

```
sudo python Robot_Car.py ↵
```

You would then simultaneously press the 1 and 2 buttons on the Wiimote to connect to the RasPi via Bluetooth. Next, you would operate the car by using the four-way switch on the Wiimote to control the direction of the car. That's it. Press the "+" button to disconnect and stop the Bluetooth session.

Future Expansions

The following would be interesting and exciting modifications to this project to expand its usefulness and hopefully the learning experience:

- Add a 5-V DC powered USB hub to the existing two USB ports that are both used. Adding a Wi-Fi adapter would provide a wireless networking capability and Internet access.
- Add a video camera, such as the GoPro Hero3. This unit has Wi-Fi connectivity, and is capable of being remotely controlled by either its own remote control accessory or an Android smartphone or tablet running the GoPro app. The remote control and Android app both have a real-time video preview function.
- Add an ultrasonic sensor that is capable of detecting objects in the forward path of the car. The Python program can easily be modified to handle sensor inputs and modify the path accordingly.
- Add an IR sensor to detect IR beams to redirect the car, basically creating an invisible fence or barrier.
- Add a GPS sensor to map the path of the car, assuming it travels more than a few meters. The Ultimate GPS receiver discussed in an earlier chapter would be an ideal unit to install on the car. The path taken could be recorded in the RasPi SD card and later sent to Google Earth for display in that application.
- Add swarm or hive behavior if you happen to have an additional car or know someone with one. The RasPi can easily handle the computing challenge that it takes to enable this type of collective behavior. The cars should be able to talk with one another using their installed adapters.
- Add an IR light and an IR sensitive camera to allow the car to maneuver in total darkness. This would be a very interesting and informative task, since rescue robots need to have this capability.

As you might imagine, many more items could be added to this list. It really is only limited by your imagination. The robot car built in this project is only a start to bigger and much better projects.

Summary

This project commenced with an overview of the car's final form. Next came specific build instructions for the chassis as well as for the "sandwich" construction that holds the various boards that make up the car.

A discussion followed regarding the servo drive motors and the associated PWM control signals. Then we looked at the procedure for setting up the RasPi to run the I²C protocol that is necessary to communicate with the 16-channel servo control board.

Some preliminary testing was next shown that proved the I²C interface functioned as expected.

A complete block diagram was shown to assist in connecting all the component boards and modules that make up the car.

After that step came a detailed procedure for setting up Bluetooth to run on the RasPi. A companion software package named *cwiid* was downloaded and installed to enable a Python program to control a Wii remote control (Wiimote).

A simple Python program was shown that allowed a user to control the car using a Wiimote. The car can go forward, turn right or left, and stop.

The project finished with many suggestions for future car expansions and modifications.

CHAPTER 14

Radon Detector

Introduction

This project could save your life. How's that for an attention grabber! Radon, as most people know is a radioactive gas that can be present in homes and that often goes undetected. This project couples a Geiger counter with the RasPi to measure and display the level of radioactivity in the monitored space. I will present plenty of background information so that you may learn precisely what radon is, how it is created, at what levels it is acceptable, and at what levels you should take some action.

The Geiger counter used in this project was built from the Velleman kit number K2645 and is shown in Fig. 14-1. This popular Geiger counter kit has been previously incorporated into other microcontroller projects, including ones using the Arduino or the Parallax Basic Stamp.

Radioactivity and Radon

Radioactivity is the process in which an unstable atom loses energy by emitting particles or rays of ionizing radiation. These particles and rays consist of:

- Alpha particles—Helium nucleus
- Beta particles—High-energy, high-speed electrons or positrons with a neutrino
- Gamma rays—Highly energetic rays

The term *ionizing radiation* relates to the upset or damage a radioactive decay particle or ray can do to a living cell. The alpha particle does the least damage because the epidermal skin layer stops it. A beta particle is a bit more penetrating, but it can be stopped by various items, including a layer of cardboard, a sheet of polyethylene plastic, aluminum foil, moderate to heavy clothing, etc. The most damaging is the gamma ray, which can penetrate deep into the human body. Gamma sources are normally shielded with lead, steel, concrete, etc.

The source of most natural radioactivity is uranium ore. It is the most abundant, naturally occurring radioactive element found in nature. The prime element found in uranium ore is ^{238}U where the 238 is the atomic weight, i.e., the number of protons and neutrons located in the nucleus. An associated key measure that will be mentioned in this discussion is *half-life* or the length of time it takes for a specific radioactive element to lose half of its activity level. For ^{238}U, the half-life is 4.5 billion years. Most other radioactive materials have a much shorter half-life, including radon, which has a half-life of 3.8 days.

Radon is a colorless, odorless gas that is one of the many products of ^{238}U radioactive decay. Radon's elemental symbol with atomic weight is ^{222}Rn. It is also one of the densest gases known. The real problem that radon poses is that it also decays and transmutes into what are known as

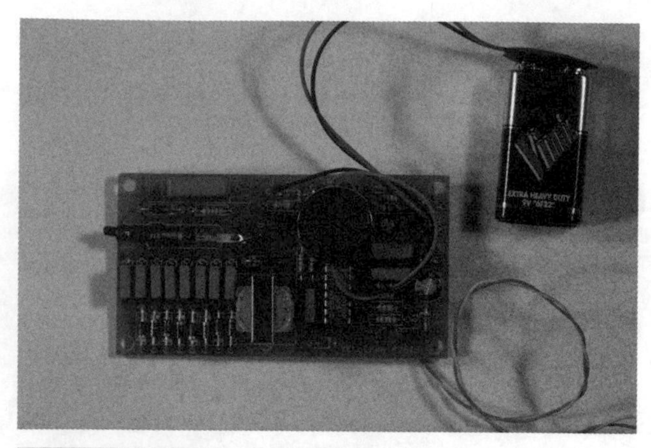

Figure 14-1 Velleman K2645 Geiger-Muller counter.

daughters: solid particles that stick to surfaces such as lung tissue.

Radon gas can and often does accumulate in confined spaces, such as basements and attics. It also occurs naturally in springs, both hot and cold. High levels of radon when inhaled over a period of time can lead to the formation of lung cancer. According to the U.S. EPA, radon is the number one cause of lung cancer in non-smokers.

Radon, being in the ^{238}U natural decay chain, is found wherever natural uranium ore is found. Often, ^{238}U is found in granite and shale deposits. I guess I am a bit unfortunate as I happen to live in New Hampshire, otherwise known as the Granite State. However, I have measured my home basement for radon and have not detected any activity. Nonetheless, there is a radon vent pipe installed in the house that vents any entrapped radon gas up and out through the roof.

Measuring Radioactivity

The *International System of Units* (SI) unit for measuring radioactivity is the *becquerel*, abbreviated as Bq. One Bq is a very tiny amount of activity. Another common unit is the *curie*, abbreviated as Ci. One Ci is many orders of magnitude greater than one Bq. The exact relation is:

$$1 \text{ Ci} = 3.7 \times 10^{10} \text{ Bq}$$

While the Bq and Ci are great for measuring radioactivity, the real question is, how much activity is dangerous to humans. Table 14-1 is a comparative listing showing radon activity levels and places where these levels might be encountered.

Dose

Absorbed dose, also known as *total ionizing dose* (TID), is a measure of the energy deposited in or affecting a medium by energetic or ionizing radiation. It is measured in joules per kilogram and represented in SI units by the *gray* (Gy). The legacy unit is the *rad*.

Table 14-1 Radon Radioactivity Levels

Bq/M³	pCi/L	Remarks
1	.027	Radon at ocean shore
10	0.27	In open air, mid-continental
100	2.70	Typical indoor domestic; average ~ 39 Bq/M³
1,000	27	Very high concentrations found in homes built on soil with high content of uranium ore present
10,000	270	Working level in a uranium mine ~ 7000 Bq/M³
100,000	2,700	Extremely high
1,000,000	27,000	Occasionally found in unventilated areas of uranium mines

Absorbed dose (now simply referred to as *dose*) is a complex measurement, because it depends on both the level of activity of the ionizing radiation and the absorbing medium, which in most cases, is human tissue.

In the United States, a commonly applied measure of dose is the *roentgen-equivalent-man* or *rem*. It is a complex, computed, and weighted average of dose representing the stochastic biological effects of ionizing radiation, which is one primary risk factor in cancer formation. The *sievert* (Sv) is the rem's equivalent international SI unit. The conversion factor is 100 rem = 1 Sv, or in more realistic units, 100 μrem = 1 μSv.

This project's radiation detector measures dose in μrem units. I have also included the following list in Table 14-2 to help put the whole concept of dose and effect into a reasonable perspective.

Converting radiation activity levels to dose is greatly simplified in this project by the use of a calibration curve supplied by the kit provider. Figure 14-2 shows the calibration curve related to the Geiger-Muller (GM) tube used in the K2645 GM counter.

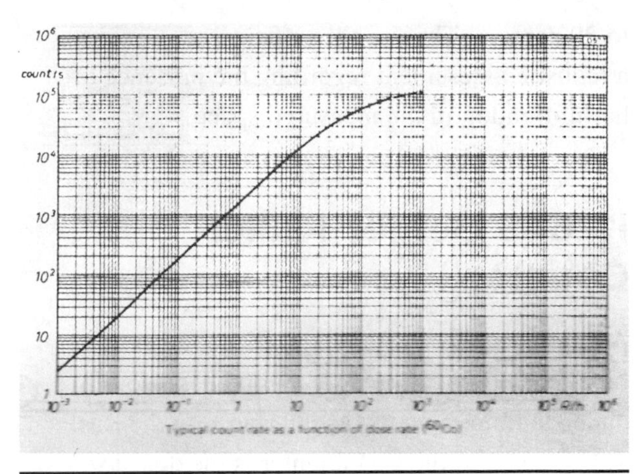

Figure 14-2 GM tube calibration curve.

The sloped portion of the curve can be easily represented by a simple equation:

$$\text{dose } (\mu\text{rem}) = (\text{c/m} \times 100)/15$$

where c/m = counts per minute

All that needs to be done is to count the number of "clicks" or triggering events that happen in a given minute and plug that number into the equation. Please note that a number of simplifications have gone into formulating this equation, and it really is accurate only to ±15

Table 14-2 Dose Examples

Dose	Effect
98 nSv	Banana equivalent dose, a whimsical unit of radiation dose
0.25 μSv	U.S. limit on effective dose from a single airport security screening
10 μSv	One set of dental radiographs
80 μSv	Average dose to people living within 16 km of Three Mile Island accident
.6 mSv	Two-view mammogram, using weighting factors; updated in 2007
5 mSv	Barium fluoroscopy
30 mSv	Single full-body CT scan
68 mSv	Estimated maximum dose to evacuees who lived closest to Fukushima
0.67 Sv	Highest dose received by a worker responding to the Fukushima emergency
6 Sv	Fatal acute doses during Goiânia accident
21 Sv	Fatal acute dose to Louis Slotin in 1946 *criticality accident* (an uncontrolled nuclear accident)
64 Sv	Nonfatal dose to Albert Stevens; spread over ~21 years, due to 1945 human experiment

to 20 percent. That is still sufficient to provide a realistic assessment regarding the presence of harmful radon in the measured space.

K2645 Geiger-Muller Counter

A brief discussion of how the GM counter functions will help provide a good basis for understanding how it was interfaced to the RasPi. The essence of the GM counter is the detector tube, which is the approximately 3-cm tube that may be seen in the upper left-hand corner of Fig. 14-1. It is a small metal cylinder typically filled with argon gas, along with a wire conductor supported lengthwise in the middle of the cylinder. The wire has an approximate 600 V DC potential on it. When a charged particle or energetic wave enters the tube, the gas breaks down or ionizes and forms a brief electrical short circuit between the inner wire and the outer metal cylinder. This causes a momentary current flow that is detected and results in the click from the counter. These clicks are counted for the dose level measurement. Figure 14-3 is the schematic for the K2645 GM counter.

There are four functional areas in the schematic:

1. A square wave oscillator that is connected to the transformer TRAF01 primary.
2. A voltage multiplier array that uses diodes and capacitors, and is connected to the TRAF01 secondary winding. This array creates the 600 V DC.
3. The GM tube.
4. A monostable oscillator that detects the momentary current pulses due to the radiation and activates a piezo electric buzzer for the audio click.

The output from the board IC labeled N6 is used as an input to the RasPi interface, which is discussed next.

GM Counter/RasPi Interface

The interface has two requirements:

1. To transform the click signals to pulses that may be counted
2. To count the pulses for preset intervals and output the resulting digital count to the RasPi

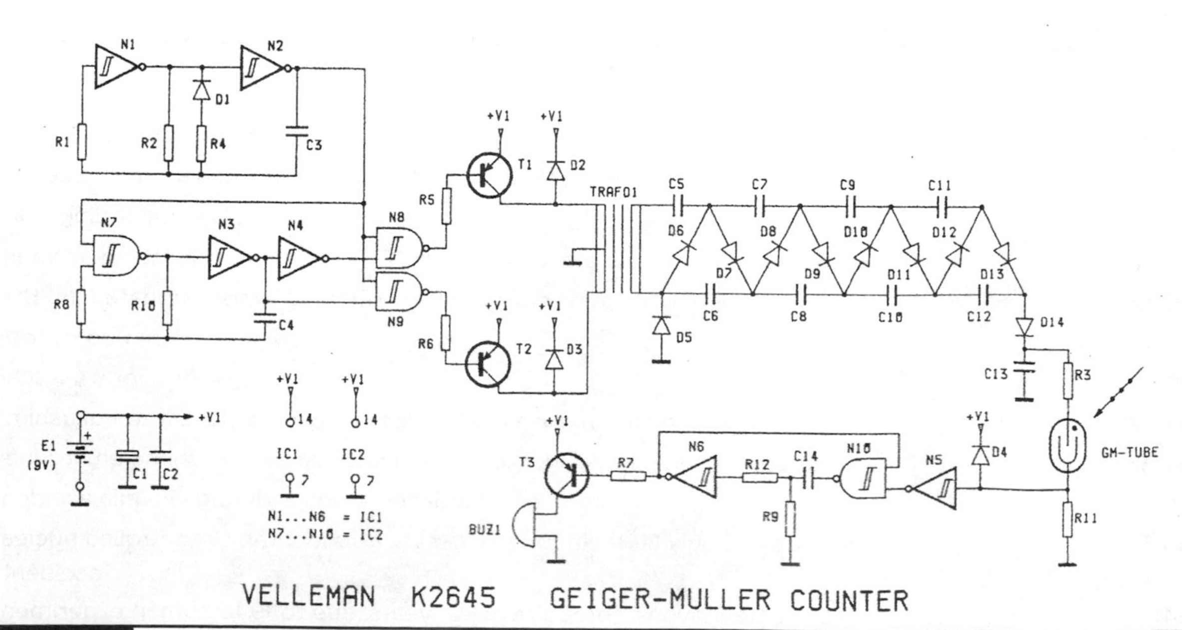

Figure 14-3 K2645 GM counter schematic.

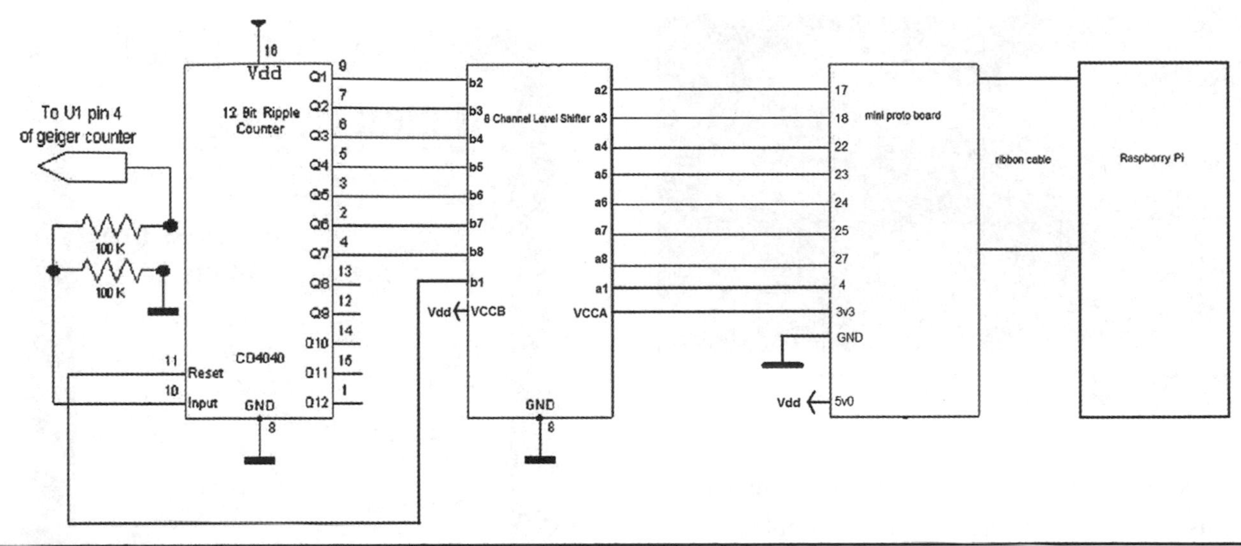

Figure 14-4 GM Counter/RasPi interface block diagram/schematic.

The block diagram/schematic for the interface is shown in Fig. 14-4.

Table 14-3 is a parts list for the interface circuit.

The voltage output from the GM counter is reduced by 50 percent and connected to the input of the CD4040 chip. Counter output bits 0 to 6 are, in turn, connected to an 8-channel level shifter chip that converts the 5-V DC counter voltages to the 3.3-V DC level, which is compatible with the RasPi GPIO input signal lines. A single GPIO output from the RasPi is connected through the level shifter chip to the counter reset line, thus allowing the ripple counter to be reset for the start of a new count period.

Figure 14-5 shows the bottom side of the GM counter with a white wire connected to the N6 output and a green wire connected to a ground point.

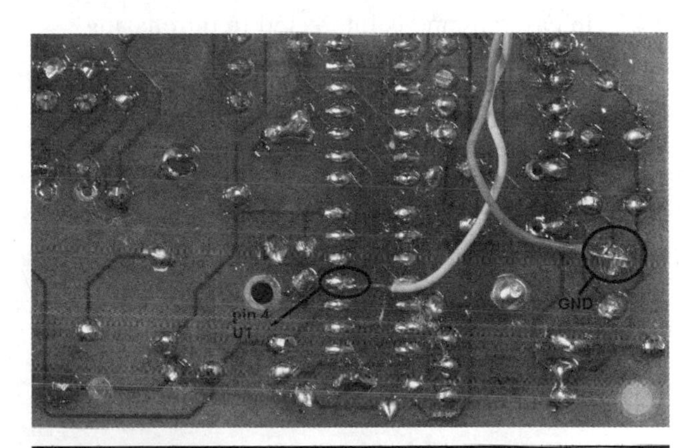

Figure 14-5 Output wire connections on the GM counter.

Initial Test Configuration

The interface circuit was first set up on a solderless breadboard with the GM counter connected, as shown in Fig. 14-6.

Table 14-3 GM Counter/RasPi Interface Parts List

Item	Distributer/Model Number	Description
Ripple counter	Digikey 296-2048-5-ND	CD4040BE CMOS 12-bit ripple counter 16 pins, DIP
Level shifter	Adafruit Industries #395	8-channel level shifter breakout board
Mini proto board	Adafruit Industries #1171	Perma-Prototype board

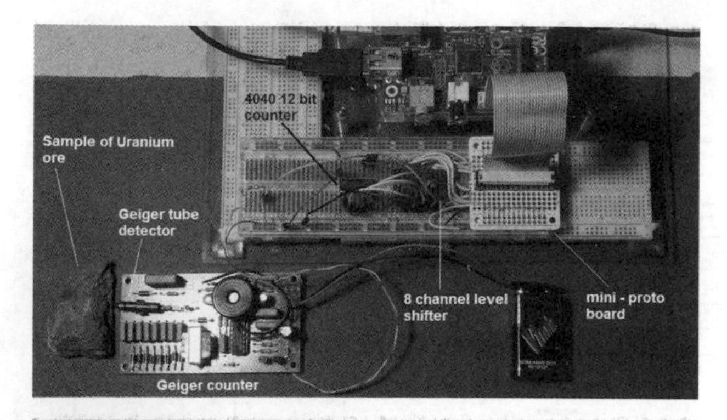

Figure 14-6 Initial test configuration.

I also purchased a sample of uranium ore that was used to test the project. The ore sample is rather innocuous, although it was certified to have a 6000 cpm—with most of that activity due to alpha radiation. The certification paper is shown in Fig. 14-7, in case anyone is interested in purchasing a sample. It is also perfectly legal to ship this type of ore sample because its activity levels fall well below any regulations governing radioactive material transport.

All eight of the RasPi GPIO pins were used in the interface. Seven were set up as inputs to read the count from the 12-bit ripple counter, and the eighth was set as an output to reset the counter.

United Nuclear

Certificate of Measured Radiation

2013

Uranium Ore Sample

MEASURED ACTIVITY:
This sample of Uranium Ore has been accurately measured for combined Beta and Gamma radiation emissions using a Stabilized Assay Meter (Eberline Instrument Corporation model #SAM2) calibrated with 1 microCi of Cs-137 on: 02 January 2013.

Emission data is measured at distance zero (direct contact), at the point of highest reading on the sample

SAFETY:
This mineral sample contains naturally occurring isotopes of Uranium, and Uranium decay products.

Direct contact with this material is not hazardous, but if the sample is crushed, or ground into a powder, ingestion or inhalation of the Uranium-laden dust may be hazardous.

Use common sense when handling this material. Wash hands after contact, do not eat or drink while handling, and avoid breathing any dust generated by the sample.

Composition & Activity Data

Sample Composition:
- [x] Carnotite (yellow)
- [x] Uraninite (black)
- [x] Gummite (brown)
- [ ] Pitchblende (black)
- [ ] Tyuyamunite (green)
- [ ] Uranophane (yellow)

Sample Location:
- [ ] Grants, NM., USA
- [x] Moab, UT., USA
- [ ] Congo, Africa

Sample Activity:

@0cm 6,000 CPM

Certified by:

United Nuclear
16429 S. Upton Rd.
Suite #1
East Lansing, Mi. 48823
www.unitednuclear.com

Figure 14-7 Uranium ore sample certification.

Below you will find the Python code, available on the book's website www.mhprofessional.com/raspi, from a program named *geiger.py* that was used to test the initial configuration:

Figure 14-8 is a screenshot of the program output captured from my laptop that was on an

geiger.py

```
import time
import RPi.GPIO as GPIO
GPIO.setmode(GPIO.BCM)
GPIO.setup(17, GPIO.IN)
GPIO.setup(18, GPIO.IN)
GPIO.setup(22, GPIO.IN)
GPIO.setup(23, GPIO.IN)
GPIO.setup(24, GPIO.IN)
GPIO.setup(25, GPIO.IN)
GPIO.setup(27, GPIO.IN)
GPIO.setup(4, GPIO.OUT)
while True:
    rdg = read_counter()
    clear_counter()
    mrem = (rdg * 100)/15
    ave = ave + mrem
```

```
    min = min + 1
    if( min == 0):
        ave = 0
    else:
        max = mrem
    print "mrem = ", mrem
    time.sleep(40)
    avemrem = ave/min
    print "average mrem = ", avemrem
    time.sleep(10)
    maxmrem = max
    print "maximum mrem = ", maxmrem
    time.sleep(10)
def read_counter():
    bit0 = GPIO.input(17)
    bit1 = GPIO.input(18)
    bit2 = GPIO.input(22)
    bit3 = GPIO.input(23)
    bit4 = GPIO.input(24)
    bit5 = GPIO.input(25)
    bit6 = GPIO.input(27)
    temp = bit0 + 2*bit1 + 4*bit2 + 8*bit3 + 16*bit4 + 32*bit5 + 64*bit6
    return temp
def clear_counter():
    GPIO.output(4, GPIO.HIGH)
    time.sleep(.5)
    GPIO.output(4, GPIO.LOW)
```

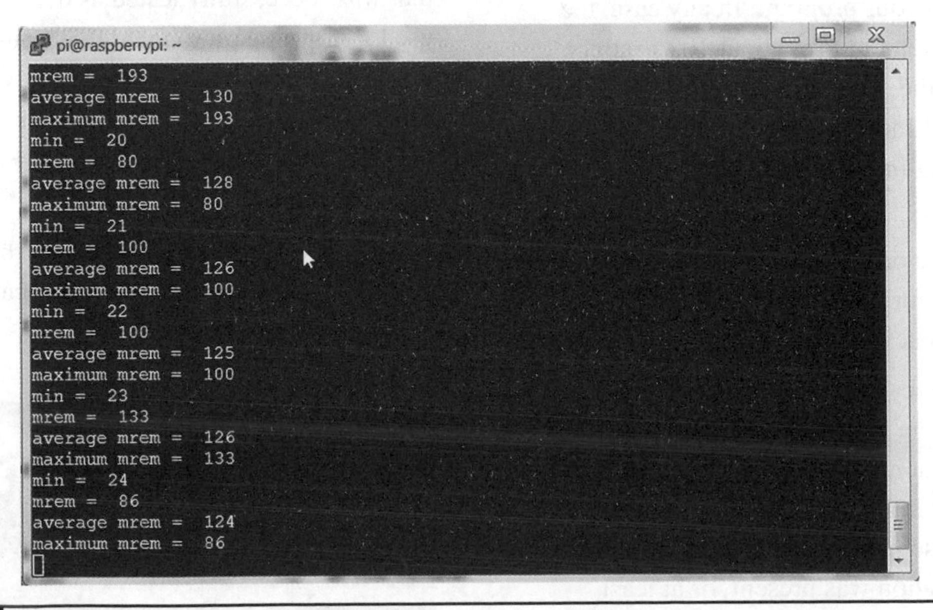

Figure 14-8 Ore sample readings.

```
pi@raspberrypi: ~
min =  7
mrem =  6
average mrem =  10
maximum mrem =  6
min =  8
mrem =  13
average mrem =  10
maximum mrem =  13
min =  9
mrem =  20
average mrem =  11
maximum mrem =  20
min =  10
mrem =  6
average mrem =  11
maximum mrem =  6
min =  11
mrem =  13
average mrem =  11
maximum mrem =  13
min =  12
mrem =  26
average mrem =  12
```

Figure 14-9 Background radiation readings.

SSH connection to the RasPi. The ore sample was put in essentially direct contact with the GM tube in an effort to obtain the highest readings possible.

It should be noted that the "mrem" shown on the screen should be interpreted as microrem not millirem. I did not have the "μ" symbol available to output. On reflection, I probably should have used "u" in lieu of "m", a minor change that you should make to your program. In any case, the dose rate is extremely low, about 100 μrem or 1 μSv—about the same as going through TSA screening four times, as shown in the dose examples listing in Table 14-2.

I also tested the area for background radiation by removing the ore sample and letting the detector simply operate for a while. Figure 14-9 shows the screenshot for background radiation.

The average reading was approximately 11 μrem, or in the banana equivalent dose range—a non-issue regarding exposure. I should also mention that the readings were taken in my basement, thus reconfirming that absolutely no radon was present, or at least detectable.

Building a Portable Radiation Detector

While the Geiger counter breadboard version worked fine, it is not a very practical solution to detecting radiation in various spaces. I thus wanted to build a portable, battery-powered version to use in the field. The first step was to assemble a prototype board for the interface circuit that was successfully tested as described above. I used a strip-line protoboard that I purchased from a local electronics supply store and wired it in exactly the same configuration as the one used for the original solderless breadboard. Figure 14-10 shows this wired board.

The only additions I made to the board are the 5-V DC voltage regulator that is located on the

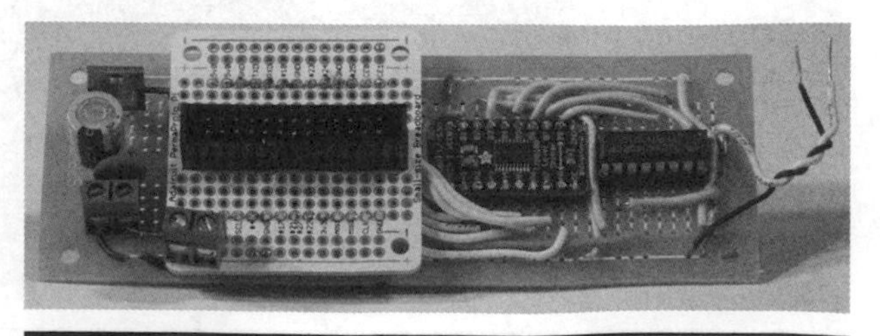

Figure 14-10 Wired interface board.

left side of the board and the two screw terminal connectors. A 12-V DC LiPo battery is connected to the leftmost terminal, while the Geiger counter is powered from the terminal connector just to the right. I must confess to a slight cheat that I made in powering the Geiger counter. It is normally powered by a 9-volt battery. All I did was put in a series resistor that dropped the LiPo's 12 V to between 9 and 9.5 V. I used a 470 Ω resistor, since the current was only a nominal 6 mA.

The black and white twisted pair of wires to the right connect to the ripple counter input and ground. Also, the mini-proto board makes the RasPi connection extremely easy; just use the ribbon cable as you did in previous projects with the Pi Cobbler.

Probably the most difficult part of the project was assembling all the pieces so that they fit into a project box. I selected a box that I thought was large enough to accommodate all the parts. It did but it was a struggle. My box is 6.25 × 4.68 × 2.25 in (15.5 × 11.5 × 6 cm), which is simply too small. I would recommend a box no smaller than 8 × 6 × 2.75 inches in order to have everything fit comfortably inside. Figures 14-11 to 14-13 are photos of the assembled portable Geiger counter not yet installed in the box.

I used a sheet of Lexan as a support plate, to which I mounted all the other components by using nylon spacers and long machine screws and nuts. This type of assembly is definitely a cut and fit

Figure 14-11 Front view of the portable Geiger counter.

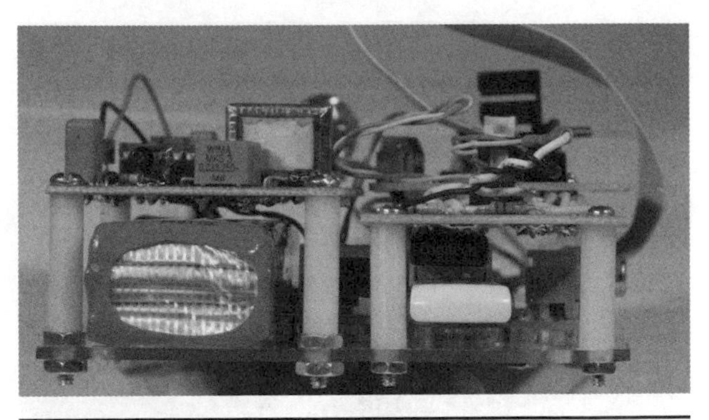

Figure 14-12 Right-side view of the portable Geiger counter.

operation where it is impossible to create a detailed plan because every installation will be slightly different. Just take your time as you carefully fit all the boards in place, and mark all the drill holes with a Sharpie prior to drilling. I must have assembled and disassembled this unit over a dozen times before finishing it.

Figure 14-14 shows the unit in the box without the top attached and with a portable analog monitor positioned behind the box.

The monitor is powered directly from the LiPo battery through a type N power jack installed in the box. Also notice that I drilled a 5/8-inch

Figure 14-13 Left-side view of the portable Geiger counter.

Figure 14-14 The Geiger counter installed in a box with the monitor.

hole through the box in order to have access to the RasPi's analog video socket. There is an on/off switch also mounted above the monitor power socket that controls the power to every component in the box.

If you look carefully at Fig.14-14, you will notice a ½-inch hole drilled through the box just to the left of the GM tube. This hole is also capped with a metal plug. I drilled this hole so that beta particles could be detected by the Geiger counter if desired. Remember from the earlier discussion that beta particles can be shielded by a thin plastic layer, so this box would certainly stop them.

Operating the Portable Geiger Counter

I installed the WiPi adapter in one of the RasPi's USB ports and put a Bluetooth adapter in the other one for the wireless keyboard and mouse unit that we also used in the robot car project. After you do this, connect the analog monitor through the side hole for the video. Don't forget to plug in the power jack. It is then a simple task of turning on the power and doing a normal boot operation. The command to start the program is:

```
sudo python geiger.py ↵
```

That should be it; you should now be seeing the radiation readings displayed on the monitor. I also connected to the Geiger counter remotely to capture a screenshot of the unit in action. Figure 14-15 shows a screenshot of the Geiger counter registering normal background readings.

```
pi@raspberrypi: ~
login as: pi
pi@192.168.1.42's password:
Linux raspberrypi 3.6.11+ #371 PREEMPT Thu Feb 7 16:31:35 GMT 2013 armv6l

The programs included with the Debian GNU/Linux system are free software;
the exact distribution terms for each program are described in the
individual files in /usr/share/doc/*/copyright.

Debian GNU/Linux comes with ABSOLUTELY NO WARRANTY, to the extent
permitted by applicable law.
Last login: Sat Mar 30 20:30:07 2013
pi@raspberrypi ~ $ sudo python geiger.py
min =  1
mrem =  0
average mrem =  0
maximum mrem =  0
min =  2
mrem =  13
average mrem =  6
maximum mrem =  13
min =  3
mrem =  13
```

Figure 14-15 Portable Geiger counter in operation.

Finally, you can connect to the Internet if you want, which makes this unit quite unique—not many network-enabled Geiger counters are available, especially at the cost of this one.

Modifications and Expansions

One modification to this project that would be useful is to attach a multi-line, I^2C, LCD display to the box and, thereby, remove the need for the monitor. You would need to make the program start automatically upon boot up, but that is relatively easy to accomplish with the many great tutorials available on the web.

I would definitely use a larger box in order to make it easier to have access to the LiPo battery. Recall the warning in the robot car chapters: *Never* use a standard automotive or even a regular power supply to recharge the LiPo battery. I just hope no one who builds this project is foolish enough to plug a wall wart into the external monitor socket in hopes of recharging the battery.

The program may also be extended to report radiation readings remotely. Right now, the program simply discards old readings, but it could be easily changed to record those readings in a log file and to send them remotely to clients using a built-in web service.

Summary

This chapter began with a discussion on radioactivity and the origins of radon gas. A distinction was made between activity levels and absorbed dose. Examples were provided to help you understand how radiation activity and dose levels are related.

A brief discussion followed on how the project's Geiger counter would function, along with a detailed explanation of how to interface the counter to the RasPi.

We then looked at an initial test configuration with sample readings taken from both background radiation and a uranium ore sample. The controlling Python program was also discussed.

Next came a detailed construction procedure on how to convert the test configuration into a portable instrument. A demonstration of this portable Geiger counter along with its networking capability followed.

Finally, I suggested some modifications and expansions to further improve the utility of this project.

CHAPTER 15

Serving Up Time with an RTC and NTP

Introduction

This project is focused on how the RasPi keeps track of time both locally and through a network. As designed, the RasPi contains no means to set or track time. This design decision was made both to keep the cost down and to minimize the printed circuit board (PCB) size. In addition, the capability would be available in the rev B model because it has a network port that is able to connect to a public *Network Time Protocol* (NTP) server. An automatic NTP connection is also built into the standard Wheezy Raspian Linux distribution designed for the RasPi.

The RasPi A models and B models without Internet access do not have public NTP service and, thus, require another means to set and track time. This is where a hardware peripheral known as a *real-time clock* (RTC) becomes very handy. First we'll discuss the RTC and then take a comprehensive look at the NTP service.

Real-Time Clock (RTC)

The RTC module, which has been available for many years, first appeared with the advent of the PC in the early 1980s. There were timing systems available before that period, but they were fairly complex and expensive devices and suitable mainly for mainframe computers. The RTC introduced with the first PC was based upon an integrated chip named the MC146818, which was manufactured at that time by the Motorola Corporation. A modern version of that chip is used in this project. It is the DS1307, pictured in Fig. 15-1, which is manufactured by the Dallas Semiconductor company, previously mentioned in Chaps. 10 and 11.

The DS1307 is an I^2C controlled device, as may be readily seen by the block diagram shown in Fig. 15-2. It is a fairly simple peripheral that uses a quartz-crystal-controlled oscillator in conjunction with a binary counter that is labeled as an *oscillator and divider* in the diagram. The crystal is rated at a nominal 32,768 Hz frequency (equivalently 32.768 kHz), which is exactly the value of 2 raised to the fifteenth power. Therefore, a 15-bit binary counter will overflow or reset once every second if a 32.768 kHz signal is input to the first stage of the counter. Theoretically, this setup will produce a once per second clock tick. Notice, I used the word theoretically, as the reality is that the clock crystals

Figure 15-1 DS1307 DIP.

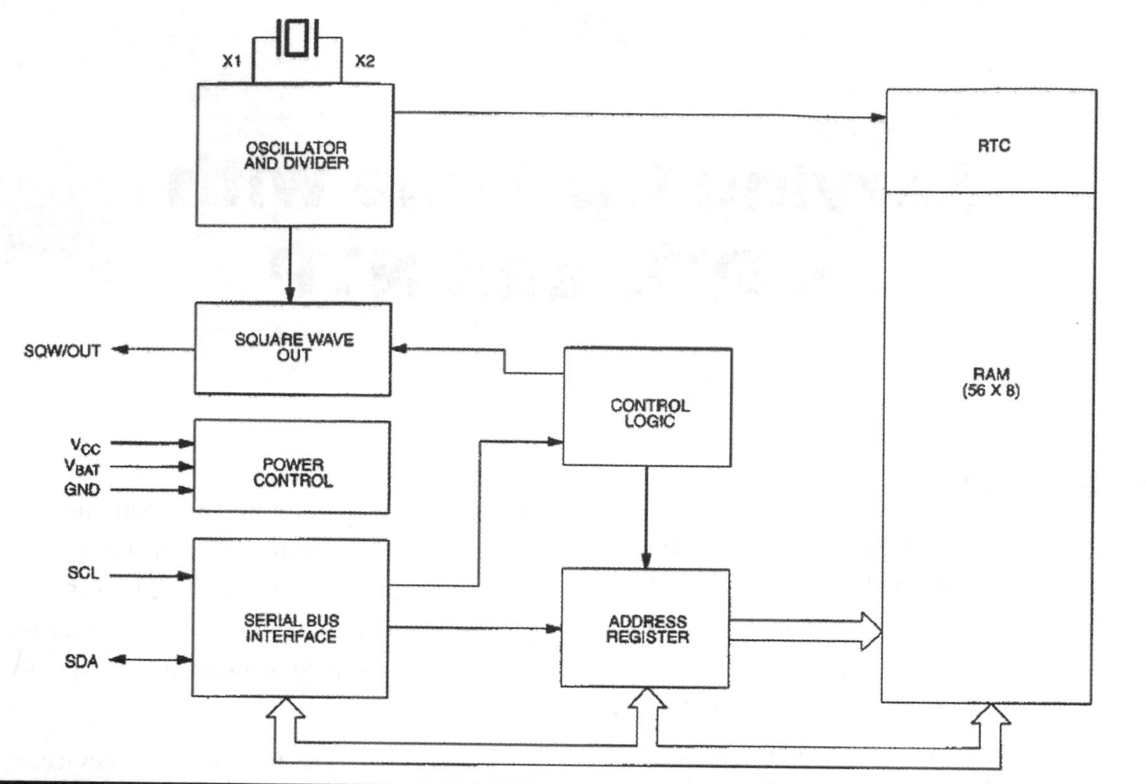

Figure 15-2 DS1307 block diagram.

typically used with the DS1307 do not produce a perfect 32.768 kHz waveform. Figure 15-3 shows the actual crystal used in this project.

You should clearly see the frequency in kHz printed on the metal can enclosing the actual piezo-electric element. These generic type RTC crystals typically will have accuracy within ±30 seconds/month when properly loaded with a matching 6 picofarad (pF) crystal and operated at a nominal 23°C ambient temperature. The operating temperature is the key parameter that affects clock accuracy, assuming the correct capacitance loading is used. Figure 15-4 shows

Figure 15-3 RTC crystal.

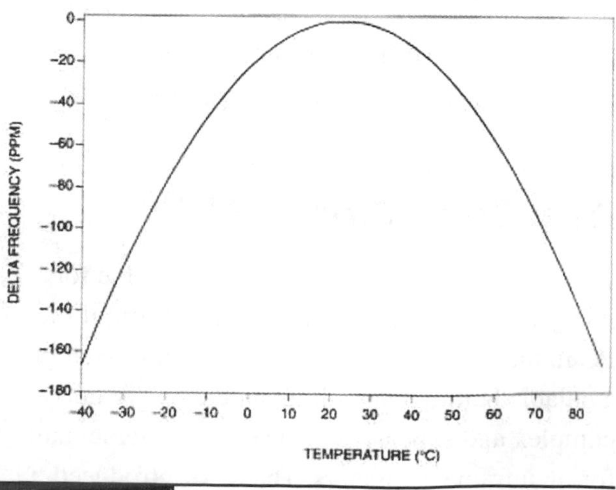

Figure 15-4 Temperature versus RTC frequency deviation.

the temperature effect as it deviates from the standard 23°C calibration point.

The ±30 seconds/month accuracy will be maintained for a RasPi operating in a home or office. However, a RasPi that operates in severe environments should expect significant reduction in RTC accuracy.

Another key feature of the RTC is the ability to maintain the time even if the host processor is turned off. This is made possible by using a long-life battery to power the oscillator and *non-volatile* (NV) RAM (write mode) in the DS1307 chip. A lithium coin cell battery is normally used for this purpose after the host is powered off. A coin cell will often last for many years before exhausting its energy. The current date and time is continually stored in the NV RAM 56-byte memory.

Figure 15-5 shows the RTC breakout board used in this project. It is built from kit number 264 purchased from Adafruit Industries. It uses the DS1307 chip along with the crystal shown in Fig. 15-3. There are two resistors not installed, but are shown as placeholders on the PCB. This is because the RasPi already has two pull-up resistors attached to the *Serial Data Line* (SDA) and *Serial Clock* (SCL) control lines, negating the need for these components.

Figure 15-5 RTC breakout board.

Table 15-1 RTC to RasPi Connections

RTC Breakout Board	Pi Cobbler Pins
5 V	5 V 0
GND	GND
SDA	SDA0
SCL	SCL0

The building kit is very easy to put together because there are only five components to install including the coin-cell holder. Each one has a placeholder noted on the PCB. Just be very careful when soldering the crystal to the board because you do not want to overheat it or excessively bend the fragile leads.

The RTC breakout board is connected to the RasPi by using a solderless breadboard with the Pi Cobbler prototype connector. The wiring connections are detailed in Table 15-1.

The physical configuration is shown in Fig. 15-6 without the ribbon cable attached to show all the jumper wires.

NOTE Even though the RTC is powered by the 5-V RasPi power supply, the I²C signal levels are still at an acceptable 3.3-V level.

Figure 15-6 RasPi and RTC setup.

RTC Software

The procedure discussed in this section is based upon a fine tutorial developed by LadyAda on the Adafruit website. The RTC chip uses the I^2C protocol for communication with the RasPi, which means that the basic I^2C software must be installed in the Wheezy Raspian distribution. The detailed I^2C installation procedure shown in Chap. 12 should be accomplished before you proceed with the following software instructions.

Enter the following command in a terminal window to confirm that the RTC is detected by the I^2C bus:

```
sudo i2cdetect –y 1 ↵
```

NOTE NOTE Replace the 1 with a 0 if a rev 1 RasPi is being used.

Figure 15-7 is a screenshot showing that the RTC board was detected at address 0 ×68.

The next step is to load the RTC module software, which is done by entering this command:

```
sudo modprobe rtc-ds1307 ↵
```

The *modprobe* application loads what is known as a *loadable kernel module* (LKM), which in this case, is named *rtc-ds1307*. You may have to run the following two commands if the LKM is not found:

```
sudo apt-get update ↵
sudo apt-get upgrade ↵
```

The next step is to instantiate the RTC object, which you must do at the root level using the following commands:

```
sudo bash ↵
echo ds1307 0x68 > /sys/class/i2c-
adapter/i2c-1/new_device ↵
exit ↵
```

NOTE i2c-1 is for a rev 2 RasPi. Use i2c-0 for a rev 1 RasPi.

You may now check the time stored in the RTC by typing the following:

```
sudo hwclock –r ↵
```

The RTC should report back, Jan 1 2000, if it has not been previously set. Figure 15-8 is a

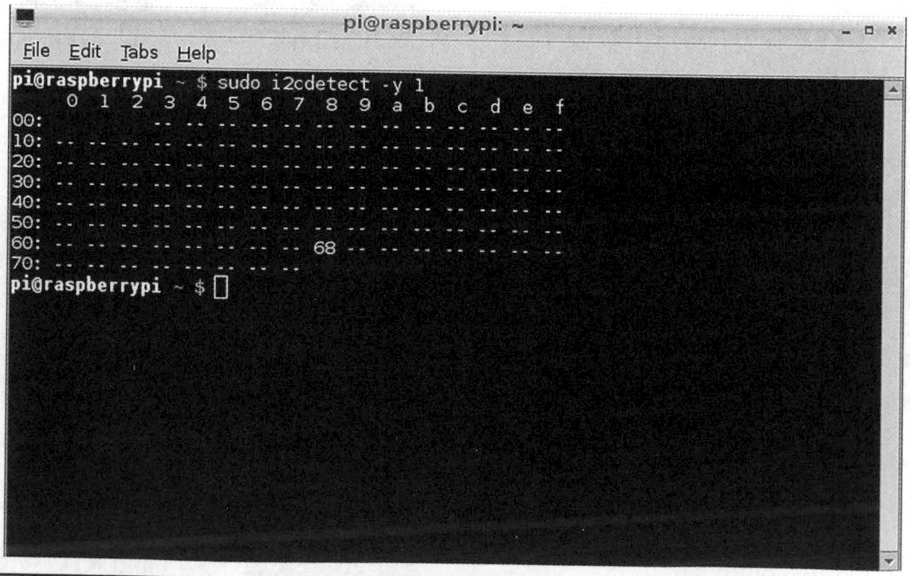

```
pi@raspberrypi: ~
File Edit Tabs Help
pi@raspberrypi ~ $ sudo i2cdetect -y 1
     0  1  2  3  4  5  6  7  8  9  a  b  c  d  e  f
00:          -- -- -- -- -- -- -- -- -- -- -- -- --
10: -- -- -- -- -- -- -- -- -- -- -- -- -- -- -- --
20: -- -- -- -- -- -- -- -- -- -- -- -- -- -- -- --
30: -- -- -- -- -- -- -- -- -- -- -- -- -- -- -- --
40: -- -- -- -- -- -- -- -- -- -- -- -- -- -- -- --
50: -- -- -- -- -- -- -- -- -- -- -- -- -- -- -- --
60: -- -- -- -- -- -- -- -- 68 -- -- -- -- -- -- --
70: -- -- -- -- -- -- -- --
pi@raspberrypi ~ $
```

Figure 15-7 Detecting the RTC board on the I^2C bus.

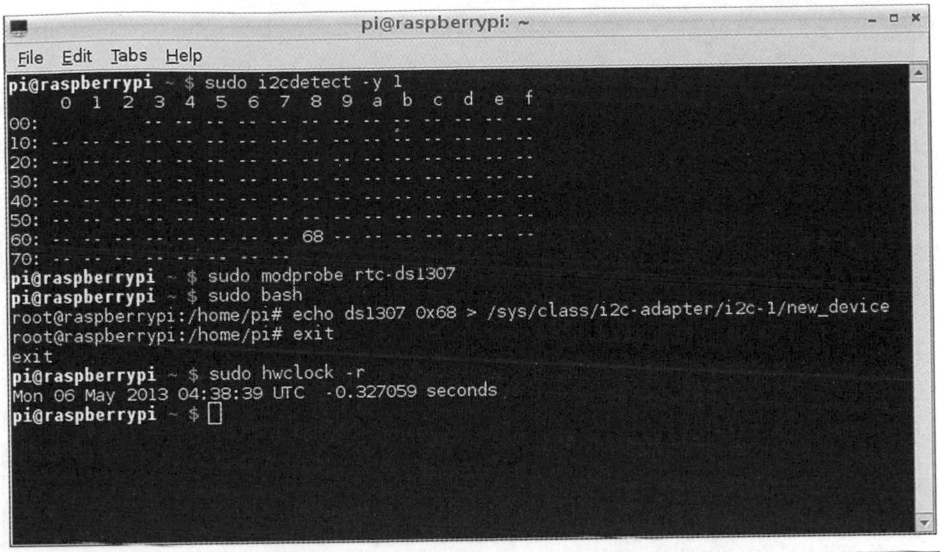

```
pi@raspberrypi: ~
File Edit Tabs Help
pi@raspberrypi ~ $ sudo i2cdetect -y 1
     0  1  2  3  4  5  6  7  8  9  a  b  c  d  e  f
00:          -- -- -- -- -- -- -- -- -- -- -- -- --
10: -- -- -- -- -- -- -- -- -- -- -- -- -- -- -- --
20: -- -- -- -- -- -- -- -- -- -- -- -- -- -- -- --
30: -- -- -- -- -- -- -- -- -- -- -- -- -- -- -- --
40: -- -- -- -- -- -- -- -- -- -- -- -- -- -- -- --
50: -- -- -- -- -- -- -- -- -- -- -- -- -- -- -- --
60: -- -- -- -- -- -- -- -- 68 -- -- -- -- -- -- --
70: -- -- -- -- -- -- -- --
pi@raspberrypi ~ $ sudo modprobe rtc-ds1307
pi@raspberrypi ~ $ sudo bash
root@raspberrypi:/home/pi# echo ds1307 0x68 > /sys/class/i2c-adapter/i2c-1/new_device
root@raspberrypi:/home/pi# exit
exit
pi@raspberrypi ~ $ sudo hwclock -r
Mon 06 May 2013 04:38:39 UTC  -0.327059 seconds
pi@raspberrypi ~ $
```

Figure 15-8 Screenshot of the RTC setup.

screenshot showing all the previous commands as well as the time. Notice that it reported a date different from Jan 1 2000 because I had rebooted the RasPi in the setup.

You will now need to set the current date and time for the RTC. This may be done in two ways. The first and probably easiest is to connect the RasPi to an Internet accessible network in order to set the current date and time. Simply plug in an Ethernet cable or a Wi-Fi adapter, and let the RasPi seek out and set its clock using the NTP service. After a minute or two, the RasPi should have acquired and set its system time to the current date and time.

The second method of setting the date and time is simply to enter it at a terminal prompt using the following sample as guidance:

```
sudo date --set="12 MAY 2013 22:15:49" ↵
```

The RTC will automatically reset its internal memory based upon your manual input.

You should always check on the system time after setting it by entering:

```
sudo date ↵
```

You may also force the RTC clock to synchronize to the system date and time by entering the following:

```
sudo hwclock —w ↵
```

Next check the time reported back by the RTC by typing in the following:

```
sudo hwclock —r ↵
```

Figure 15-9 is a screenshot showing the set time command as well as the time reported back by the RTC.

The next portion of this procedure concerns how to permanently configure the RasPi to use the RTC board. The rtc-ds1307 module must first be added to the LKM list that is stored in the /etc/modules file. To do this, edit the modules file by adding a line, "rtc-ds1307," at the end of the list. Figure 15-10 shows this edit using the nano editor.

The last remaining step is to edit a file named *rc.local* that is located in the/etc directory. This file contains scripts that are run at the end of the boot process, which is how the RTC object will be created once the following script lines are entered into the rc.local file:

```
pi@raspberrypi ~ $ sudo i2cdetect -y 1
     0  1  2  3  4  5  6  7  8  9  a  b  c  d  e  f
00:          -- -- -- -- -- -- -- -- -- -- -- -- --
10: -- -- -- -- -- -- -- -- -- -- -- -- -- -- -- --
20: -- -- -- -- -- -- -- -- -- -- -- -- -- -- -- --
30: -- -- -- -- -- -- -- -- -- -- -- -- -- -- -- --
40: -- -- -- -- -- -- -- -- -- -- -- -- -- -- -- --
50: -- -- -- -- -- -- -- -- -- -- -- -- -- -- -- --
60: -- -- -- -- -- -- -- -- 68 -- -- -- -- -- -- --
70: -- -- -- -- -- -- -- --
pi@raspberrypi ~ $ sudo modprobe rtc-ds1307
pi@raspberrypi ~ $ sudo bash
root@raspberrypi:/home/pi# echo ds1307 0x68 > /sys/class/i2c-adapter/i2c-1/new_device
root@raspberrypi:/home/pi# exit
exit
pi@raspberrypi ~ $ sudo hwclock -r
Mon 06 May 2013 04:38:39 UTC  -0.327059 seconds
pi@raspberrypi ~ $ sudo hwclock -w
pi@raspberrypi ~ $ sudo hwclock -r
Sun 12 May 2013 02:03:11 UTC  -0.253554 seconds
pi@raspberrypi ~ $
```

Figure 15-9 Screenshot of setting and checking the RTC date and time.

```
sudo nano /etc/rc.local ↵
```

```
echo ds1307 0x68 > /sys/class/i2c-
adapter/i2c-1/new_device ↵
sudo hwclock –s ↵
```

NOTE i2c-1 is for a rev 2 RasPi. Use i2c-0 for a rev 1 RasPi.

Figure 15-11 shows this edit using the nano editor.

I shut down the RasPi after completing the previous configuration steps and waited until the next day to restart the system in order to check the RTC function. Figure 15-12 shows that the date function worked perfectly, indicating that the RTC was performing as expected and accurately tracking the passage of time.

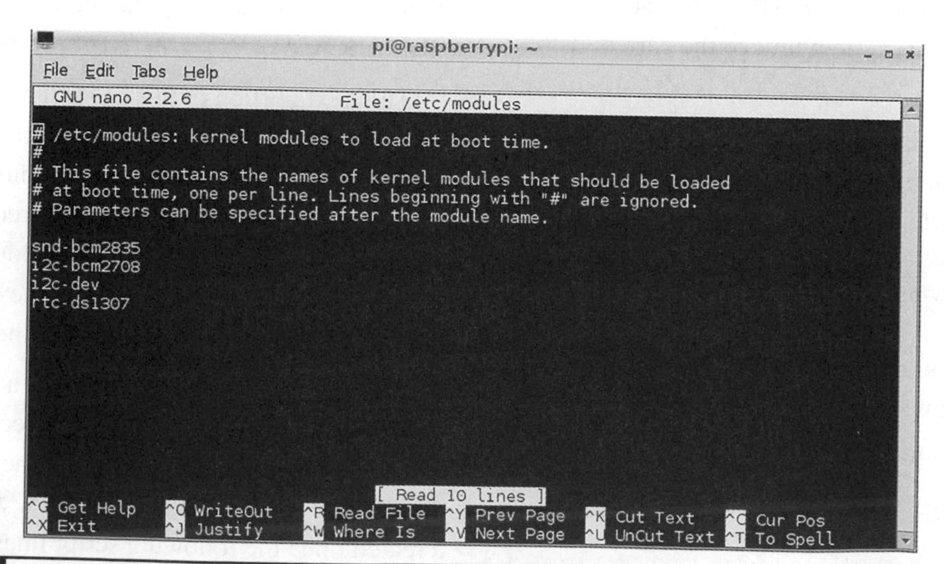

Figure 15-10 LKM module screenshot.

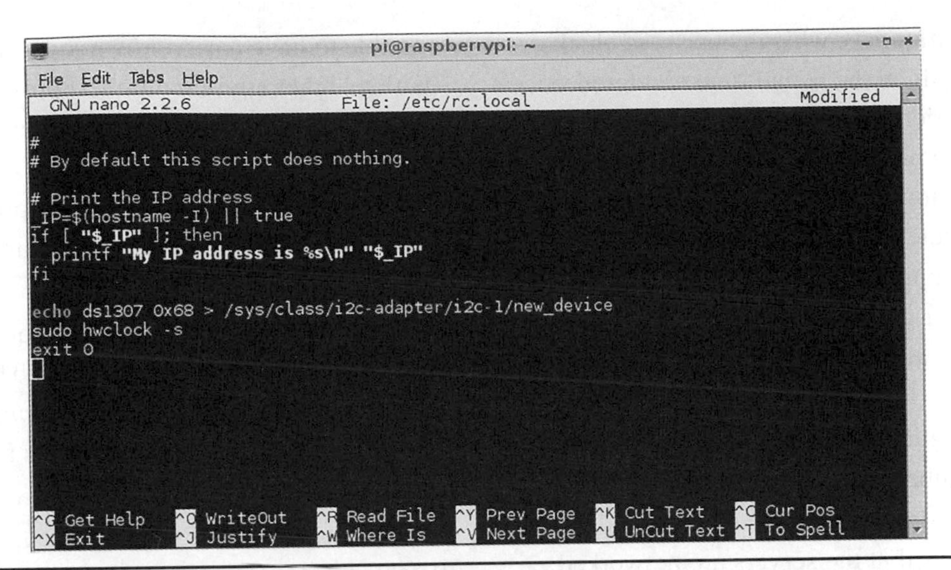

Figure 15-11 The edited rc.local file screenshot.

Figure 15-12 Screenshot of the date function.

Introduction to the Network Time Protocol (NTP)

NTP is probably the most common way time is set and maintained in today's modern networks, including the Internet. The protocol itself was created by Dr. David Mills at the University of Delaware in 1985. Currently, Dr. Mills is an emeritus professor at the University of Delaware, where he was a full professor from 1986 to 2008. He, along with a team of volunteers, is still involved with the development and maintenance of NTP.

In reality, the NTP is implemented by a layered hierarchical network of computers, each one of which is set up as a time server running the NTP protocol. Figure 15-13 is a representative block diagram of this hierarchy.

The layers in the NTP hierarchy are referred to as strata, starting with 0 and progressing to as high as 256. In reality, layers 16 or higher are considered unsynchronized and are probably not implemented. The highest layer, shown in

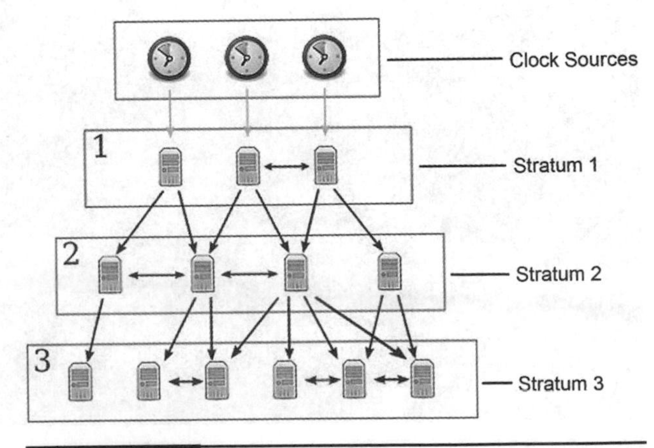

Figure 15-13 NTP server hierarchy.

the figure, is stratum 0, which consists of clock sources from which the actual time is referenced. There are a variety of primary clock sources available in the public NTP network, including sophisticated atomic clocks, GPS clocks, and the National Institute of Standards and Technology (NIST) time signal radio station WWVB to name a few. Stratum 1 connects directly to the clock sources and usually is a high-quality clock server source, but this is not always the case. NTP does not inherently guarantee that a server situated in the lowest-numbered stratum will provide the highest quality or most reliable clock signal. The reason for this is that the servers are networked and are constantly crosschecking the clock signals from other servers, both in their same stratum as well as in the strata situated logically above them. Poor-quality clock signals will be rejected from adjoining servers, and only quality signals will be passed on to other servers. Thus, it is entirely possible to have a server on a lower-level stratum (with a higher stratum number) provide a higher-quality clock signal than a server on a stratum closer to the primary clock sources. Don't worry if all this sounds a bit confusing; I provide it only as background to promote your awareness of the underlying NTP structure.

A series of public NTP servers connected to the Internet are known as *pool servers*. The URLs for these servers are stored in the *ntp.conf* file located in the /etc directory. Figure 15-14 is a screenshot of this file.

There are four NTP servers listed, starting with 0.debian.pool.ntp.org through 3.debian.pool.ntp.org. The RasPi will typically connect to one or more of these servers each time it is booted. I am unsure whether the lead number 0 to 3 represents a stratum level; however, it is irrelevant, as the NTP software will automatically select the best clock signal and use it.

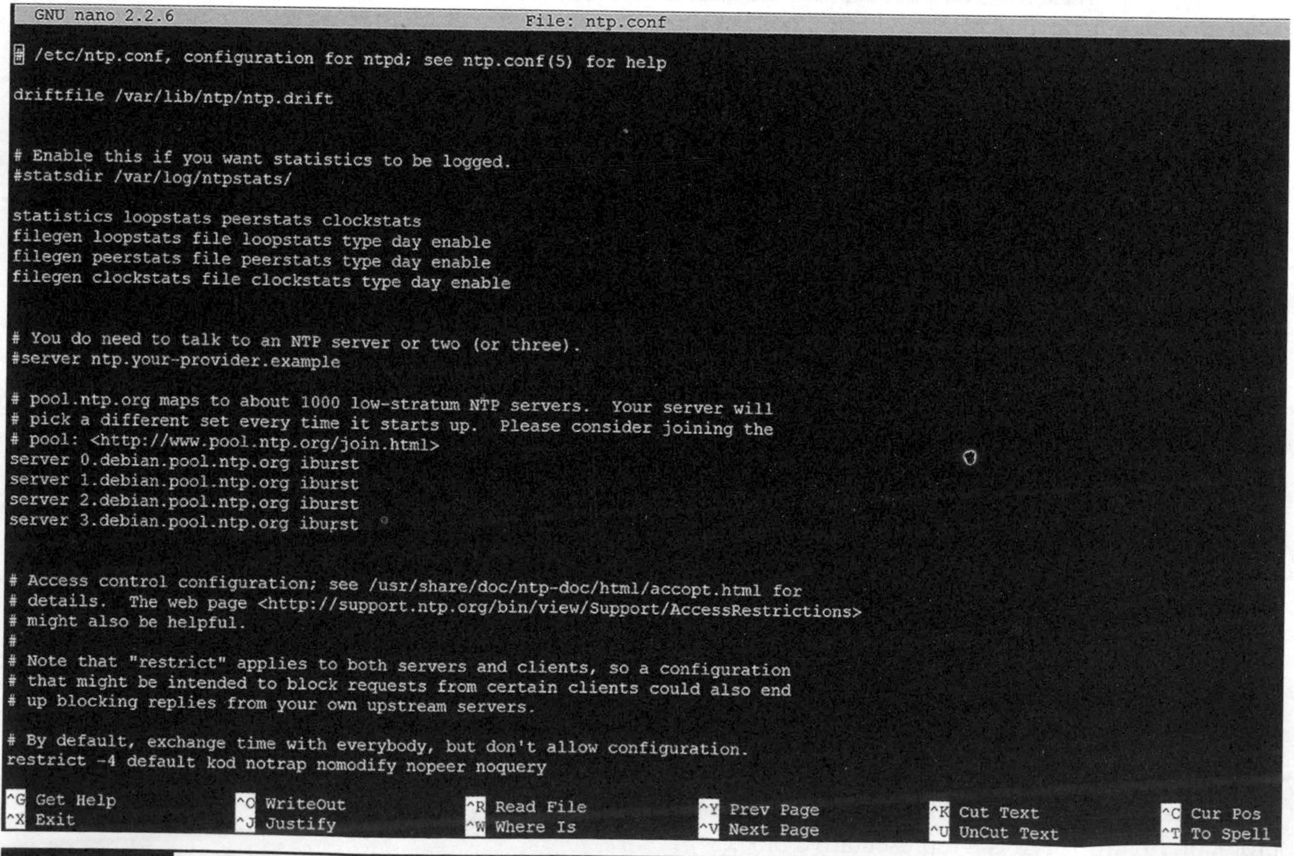

```
GNU nano 2.2.6                    File: ntp.conf

# /etc/ntp.conf, configuration for ntpd; see ntp.conf(5) for help

driftfile /var/lib/ntp/ntp.drift


# Enable this if you want statistics to be logged.
#statsdir /var/log/ntpstats/

statistics loopstats peerstats clockstats
filegen loopstats file loopstats type day enable
filegen peerstats file peerstats type day enable
filegen clockstats file clockstats type day enable


# You do need to talk to an NTP server or two (or three).
#server ntp.your-provider.example

# pool.ntp.org maps to about 1000 low-stratum NTP servers.  Your server will
# pick a different set every time it starts up.  Please consider joining the
# pool: <http://www.pool.ntp.org/join.html>
server 0.debian.pool.ntp.org iburst
server 1.debian.pool.ntp.org iburst
server 2.debian.pool.ntp.org iburst
server 3.debian.pool.ntp.org iburst


# Access control configuration; see /usr/share/doc/ntp-doc/html/accopt.html for
# details.  The web page <http://support.ntp.org/bin/view/Support/AccessRestrictions>
# might also be helpful.
#
# Note that "restrict" applies to both servers and clients, so a configuration
# that might be intended to block requests from certain clients could also end
# up blocking replies from your own upstream servers.

# By default, exchange time with everybody, but don't allow configuration.
restrict -4 default kod notrap nomodify nopeer noquery

^G Get Help    ^O WriteOut    ^R Read File   ^Y Prev Page   ^K Cut Text     ^C Cur Pos
^X Exit        ^J Justify     ^W Where Is    ^V Next Page   ^U UnCut Text   ^T To Spell
```

Figure 15-14 Screenshot of the ntp.conf file.

The RasPi NTP software is run as a daemon and requires no manual intervention for normal operation. The software is formally named *ntpd* with the "d" indicating a daemon application.

Building a RasPi NTP Server

I will show you how to build an NTP server that can provide accurate clock signals without the need to connect to one of the pool servers described above. Sometimes it is necessary to provide an independent NTP network server that does not rely on an Internet connection and can serve as a central clock source for all networked computers. There are commercial NTP servers available, but they typically cost anywhere from $1500 to $2000. Using a RasPi with a GPS will drop the cost to less than $100 and will provide the desired functionality.

The GPS used in this stage of the project is the same model previously discussed in Chap. 5. I will not repeat all the background and setup information provided in that chapter and will simply assume that you will follow those instructions in establishing the GPS to RasPi UART communications link. Refer to Fig. 5-15 in Chapter 5 to see the essential wiring required to connect the GPS module and the RasPi. You should ensure that the gpsd package is installed on the RasPi, as you will need it to process the GPS clock data. I recommend that you also install the CuteCom terminal control program. Using that program will enable you to easily confirm that both the UART link and the GPS module are functioning properly. Again, simply follow the procedures detailed in Chap. 5 to set up the GPS module and the RasPi to use the UART communications device ttyAMA0.

The ntp.conf file located in the etc directory must be edited to force the ntpd application to use the GPS instead of one of the pool servers. I commented out all the pool servers and added the following two lines to that section of the file:

```
server 127.127.28.0 minpoll 4
fudge 127.127.28.0 time1 0.183 refid
NMEA
```

Figure 15-15 is a screenshot of the edited ntp.conf file.

NOTE **The pool servers will be uncommented after the GPS time server functionality has been proven, since it is good practice to leave these servers available.**

Next the following series of commands will set up and run the GPS module as a clock source for the RasPi. Ensure that no Wi-Fi adapter or Ethernet cable is attached because they could provide a clock source if you didn't comment out the pool servers listed in the ntp.conf file. All the commands should be entered in the sequence shown:

```
sudo killall gpsd ↵
sudo gpsd /dev/ttyAMA0 -F /var/run/
gpsd.sock ↵
sudo service ntp restart ↵
```

The `killall gpsd` command stops the gpsd daemon so that a logical socket link may be established by the next command. The `/dev/ttyAMA0` parameter in the second command specifies that the GPS data will be sourced from the UART. The last two parameters, `-F /var/run/gpsd.sock`, set up a control socket for device adds and removals with the full path description to the socket. The last command starts the ntp daemon, at which point it will attempt to retrieve the clock signal from the source specified in the ntp.conf file.

The GPS module will continue to blink its LED until a solid lock is obtained on four or more satellites. A reliable time is not available until the LED stops blinking, which could take several minutes depending on the antenna signal strength. I used an external antenna and was able

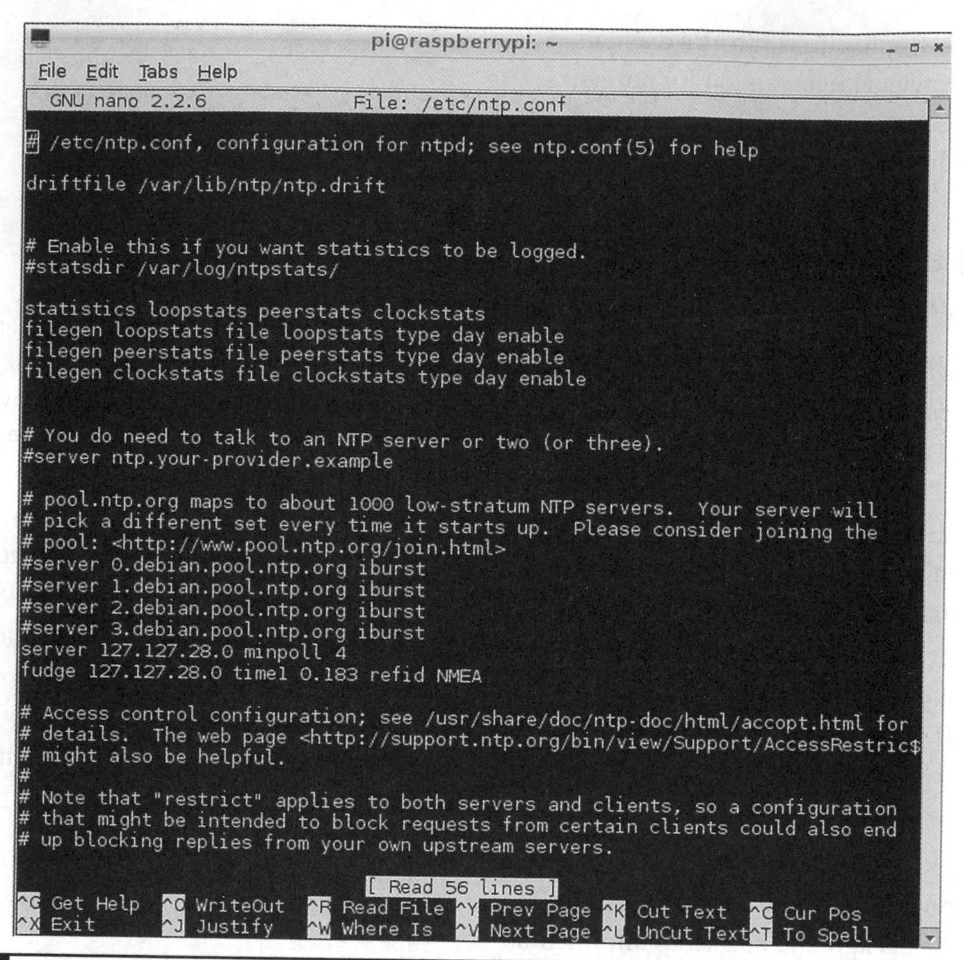

Figure 15-15 Screenshot of the ntp.conf file.

to consistently obtain a good lock in about two minutes. You may now use the date command to check on the precision time that is obtained from the GPS system.

The ntp.conf file must now be edited if all is well at this stage in the process. The following line indicates that the RasPi should be available as a stratum 10 server to other networked computers if desired:

```
server 127.127.28.0 stratum 10
```

Additionally, uncomment and edit the following existing line in ntp.conf:

```
#broadcast 192.168.123.255
```

I changed this line to:

```
broadcast 192.168.1.255
```

This line allows the NTP server to broadcast the time to all computers located on the same subnet as the server. Your subnet may be slightly different. Type these three commands to ensure that the NTP server is operating:

```
sudo killall gpsd ↵
sudo gpsd /dev/ttyAMA0 –F /var/run/
gpsd.sock ↵
sudo service ntp restart ↵
```

I recommend using a program named *ntpdate* to confirm that a client computer is connected to and using the NTP server. Install the ntpdate program by typing:

```
sudo apt-get install ntpdate ↵
```

Type the following command on the client computer (another RasPi) that is intended to use the NTP clock signal to confirm it is being received:

```
sudo ntpdate –vd 192.168.1.43 ↵
```

The local IP address for my NTP server is 192.168.1.43. Figure 15-16 is a screenshot of the ntpdate program output, clearly showing that the NTP service is being provided by the RasPi NTP server.

Radio Receiver Clock Source

Originally there was an additional section to this chapter that dealt with a radio receiver clock source, which was fixed tuned to WWVB, the U.S. synchronized time source located in Ft. Collins, CO. WWVB transmits a digital clock signal on 60 kHz 24 hours per day, seven days a week. There are several low-cost receiver kits available that can receive WWVB throughout the lower 48 states. I used a $15 kit that is available from PV Electronics, which is actually situated in the UK. I stated all of the above because even though the receiver functioned properly, the RasPi itself generated so much radio frequency interference (RFI) that it literally swamped out the WWVB signal. The RasPi is FCC certified to be compliant with class B, low-power consumer devices; however, it is not required to be tested for signal emission interference (EMI) at levels lower than 450 kHz. Obviously, 60 kHz is much lower than the minimum EMI requirement, and thus, it acts as an inadvertent interference, while being perfectly compliant with all FCC regulations for this device type. The conclusion is not to use the RasPi as a WWVB controller unless you are willing to experiment with various shielding techniques to mitigate the inadvertent interference, which I did without a lot of success.

```
pi@raspberrypi: ~
File Edit Tabs Help
pi@raspberrypi ~ $ sudo ntpdate -vd 192.168.1.43
13 May 21:58:33 ntpdate[3268]: ntpdate 4.2.6p5@1.2349-o Sat May 12 10:38:22 UTC
2012 (1)
transmit(192.168.1.43)
receive(192.168.1.43)
transmit(192.168.1.43)
receive(192.168.1.43)
transmit(192.168.1.43)
receive(192.168.1.43)
transmit(192.168.1.43)
receive(192.168.1.43)
server 192.168.1.43, port 123
stratum 2, precision -20, leap 00, trust 000
refid [192.168.1.43], delay 0.02670, dispersion 0.00011
transmitted 4, in filter 4
reference time:    d53c1895.06e8c193  Mon, May 13 2013 21:57:41.026
originate timestamp: d53c18cf.dde4a91c  Mon, May 13 2013 21:58:39.866
transmit timestamp:  d53c18cf.e40fe6b0  Mon, May 13 2013 21:58:39.890
filter delay:  0.02785  0.02676  0.02670  0.02678
         0.00000  0.00000  0.00000  0.00000
filter offset: -0.02480 -0.02518 -0.02507 -0.02493
         0.000000 0.000000 0.000000 0.000000
delay 0.02670, dispersion 0.00011
offset -0.025077

13 May 21:58:39 ntpdate[3268]: adjust time server 192.168.1.43 offset -0.025077
sec
pi@raspberrypi ~ $
```

Figure 15-16 The ntpdate program output.

Summary

The first part of the chapter covered how to set up a hardware clock to be used with the RasPi because it does not incorporate one and would lack a time capability if not connected to the Internet. The hardware clock was based on the DS1307 chip and used the I^{2}C bus to communicate with the RasPi. The clock board also incorporated a lithium coin cell battery that maintained the time even if the RasPi was turned off.

Next, I presented a brief background about the origin and function of the Network Time Protocol (NTP) that the RasPi uses when connected to the Internet. The NTP is implemented on the RasPi by the ntpd daemon, which is configured by a file named ntp.conf located in the etc directory.

Finally, I presented a section on how to create your own NTP server based upon a GPS clock source. The GPS module, first shown in Chap. 5, was the source used. It communicated with the RasPi using the UART link. Several changes had to be made to the ntp.conf file to enable the NTP server function. I finished the section discussion by demonstrating how to install and use the ntpdate application to test and prove that a client computer on the local network was actually using the RasPi NTP server.

Index

G

H

R

S